Physics Research and Technology

Leptons

Classes, Properties and Interactions

PHYSICS RESEARCH AND TECHNOLOGY

Additional books and e-books in this series can be found on Nova's website under the Series tab.

Physics Research and Technology

Leptons

Classes, Properties and Interactions

Christopher M. Villegas
Editor

Copyright © 2019 by Nova Science Publishers, Inc.

All rights reserved. No part of this book may be reproduced, stored in a retrieval system or transmitted in any form or by any means: electronic, electrostatic, magnetic, tape, mechanical photocopying, recording or otherwise without the written permission of the Publisher.

We have partnered with Copyright Clearance Center to make it easy for you to obtain permissions to reuse content from this publication. Simply navigate to this publication's page on Nova's website and locate the "Get Permission" button below the title description. This button is linked directly to the title's permission page on copyright.com. Alternatively, you can visit copyright.com and search by title, ISBN, or ISSN.

For further questions about using the service on copyright.com, please contact:
Copyright Clearance Center
Phone: +1-(978) 750-8400   Fax: +1-(978) 750-4470   E-mail: info@copyright.com.

NOTICE TO THE READER

The Publisher has taken reasonable care in the preparation of this book, but makes no expressed or implied warranty of any kind and assumes no responsibility for any errors or omissions. No liability is assumed for incidental or consequential damages in connection with or arising out of information contained in this book. The Publisher shall not be liable for any special, consequential, or exemplary damages resulting, in whole or in part, from the readers' use of, or reliance upon, this material. Any parts of this book based on government reports are so indicated and copyright is claimed for those parts to the extent applicable to compilations of such works.

Independent verification should be sought for any data, advice or recommendations contained in this book. In addition, no responsibility is assumed by the publisher for any injury and/or damage to persons or property arising from any methods, products, instructions, ideas or otherwise contained in this publication.

This publication is designed to provide accurate and authoritative information with regard to the subject matter covered herein. It is sold with the clear understanding that the Publisher is not engaged in rendering legal or any other professional services. If legal or any other expert assistance is required, the services of a competent person should be sought. FROM A DECLARATION OF PARTICIPANTS JOINTLY ADOPTED BY A COMMITTEE OF THE AMERICAN BAR ASSOCIATION AND A COMMITTEE OF PUBLISHERS.

Additional color graphics may be available in the e-book version of this book.

Library of Congress Cataloging-in-Publication Data

ISBN: 978-1-53614-929-6

Published by Nova Science Publishers, Inc. † New York

CONTENTS

Preface		vii
Chapter 1	Elementary Particle Masses: An Alternative to the Higgs Field *Raymond Fèvre*	1
Chapter 2	The Origin of SU (5) Symmetry by Lepton Electric Charge Swap Quaternions Algebra *E. Koorambas*	27
Chapter 3	Scattering $v_{\mu,\tau} - e^-$ in the 331RHv Model and Electromagnetic Properties *A. Gutiérrez-Rodríguez, A. Burnett-Aguilar and M. A. Hernández-Ruíz*	57
Chapter 4	Tau-Lepton: Dipole Moments in the B-L Model *A. Gutiérrez-Rodríguez and M. A. Hernández-Ruíz*	69
Bibliography		79
Related Nova Publications		113
Index		117

PREFACE

In *Leptons: Classes, Properties and Interactions*, the authors develop a model to calculate the masses of charged leptons by quantifying the electrostatic field generated by these particles. This model is extended to weak and strong interactions in order to calculate the masses of all elementary fermions.

Next, by taking the SU(2) group of weak interactions in the presence of Electric Charge Swap-symmetry as a starting point, this compilation show that ordinary and non-regular leptons are related by the ECS-rotational SO(3) group. By considering the ECS-Hamiltonian quaternions for leptons, the authors find that the SU(5) Grand Unified Theory originates from the image of normalized quaternions group N(Q8) under the ECS-rotations.

The differential cross-section corresponding to a dispersion process in the context of the 331RHv model is calculated in order to use the results in terrestrial and astrophysical experiments. The differential cross-section is written in terms of the mass of the new gauge boson Z', the mixing angle, the magnetic moment of the neutrino, and the charge radius.

In closing, the authors obtain analytical expressions for the total cross section of the process $e^+e- \to \tau^+\tau^-\gamma$ in the context of the B-L model. The total cross section is analyzed in terms of the mass of the new gauge boson

Z', the mixing angle θ' of the B-L model, the magnetic moment a_τ and the electric dipole moment d_τ of the τ-lepton.

Chapter 1 - In a recent paper, the authors developed a model to calculate the masses of charged leptons by quantifying the electrostatic field generated by these particles. In the present article, the authors extend this model to weak and strong interactions in order to calculate the masses of all elementary fermions. An extension of the model concerned with the masses of hypothetical sterile neutrinos (including dark matter) is also attempted. In this way, the model could be an alternative to the BEH mechanism of the standard model.

Chapter 2 - In this chapter, taking the SU(2) group of weak interactions in the presence of Electric Charge Swap (ECS)-symmetry as a starting point, the authors show that ordinary and non-regular (ECS) leptons are related by the ECS-rotational SO(3) group. By considering the ECS-Hamiltonian quaternions for leptons, the authors find that the SU(5) Grand Unified Theory (GUT) originates from the image of normalized quaternions group $N(Q_8)$ under the ECS-rotations. Furthermore, SU(5) symmetry as well as its Standard Model (SM) subgroup are not fundamental symmetries, since they can be derived by ECS-Lepontic quaternions. This means that gluons and photons are not fundamental particles of nature.

Chapter 3 - The authors calculate the differential cross-section corresponding to the dispersion process $\nu_{\mu,\tau} - e^-$ in the context of the 331RHν model in order to use our results in terrestrial and astrophysical experiments. The differential cross-section is written in terms of the mass of the new gauge boson Z', the mixing angle ϕ, the magnetic moment of the neutrino μ_ν, and the charge radius $\langle r^2 \rangle$. Furthermore, our results are compared with those corresponding to the Standard Model in the decoupling limit $M_{Z'} \to \infty$ and $\phi = 0$.

Chapter 4 - The authors obtain analytical expressions for the total cross section of the process $e^+e^- \to \tau^+\tau^-\gamma$ in the context of the B-L model. The total cross section is analyzed in terms of the mass of the new gauge boson Z', the mixing angle θ' of the B-L model, the magnetic moment a_τ and the electric dipole moment d_τ of the τ-lepton.

In: Leptons
Editor: Christopher M. Villegas

ISBN: 978-1-53614-929-6
© 2019 Nova Science Publishers, Inc.

Chapter 1

ELEMENTARY PARTICLE MASSES: AN ALTERNATIVE TO THE HIGGS FIELD

Raymond Fèvre[*]
Dijon, France

ABSTRACT

In a recent paper [1], we developed a model to calculate the masses of charged leptons by quantifying the electrostatic field generated by these particles. In the present article, we extend this model to weak and strong interactions in order to calculate the masses of all elementary fermions. An extension of the model concerned with the masses of hypothetical sterile neutrinos (including dark matter) is also attempted. In this way, the model could be an alternative to the BEH mechanism of the standard model.

Keywords: elementary fermions masses, Higgs Field

PACS: 12.15.Ff _ 14.60.Pq

---

[*] Email: clr.fevre@wanadoo.fr.

1. Introduction

The BEH mechanism, called the "Higgs Field", was conceived in the 1960s to overcome inconsistencies in the standard model. The standard model states that the masses of the elementary particles are all null.

A quantum field was therefore introduced in order to impart a mass to the elementary particles, first for the bosons W and Z, and then for the elementary fermions.

A particle must be associated with this quantum field. The particle theorists have thus assumed the existence of a Higgs boson associated with the Higgs field.

The experimental discovery in 2012 at CERN of a boson with a mass close to 125 GeV was presented almost unanimously as corresponding to the Higgs boson. The measurements made next have shown that the properties of this particle correspond to the BEH model, in particular that the coupling constants are proportional to the particles masses.

The BEH model seems therefore robust. But it does not predict the mass of the Higgs boson. Moreover, it does not allow the calculation of elementary particles masses, or even to confirm that neutrinos have a mass. It is therefore justified to look for a model that allows the effective calculation of these masses.

The author of the present chapter published a paper in JFAP on the subject in 2016, suggesting an alternative for the boson using particles composed of ultra-relativistic tau leptons and bottom quarks [2]. Furthermore, in 2014 we published an article in Physics Essays [1] suggesting an alternative to the BEH mechanism in order to explain and calculate the mass of charged leptons. The model infers the masses of these leptons from the quantization of the electrostatic field generated by the particles.

This chapter resumes, in simpler form, the model used in the article cited above. To simplify the presentation of the model, the term corresponding to radiative corrections of Coulomb's law (called RC below) has been omitted. The calculations corresponding to this term are given in the appendix.

The model is then extended to calculate the masses of all the elementary fermions.

2. THE QUANTUM MODEL

Each of the three charged leptons (electron, muon, tau) creates an electrostatic field. The expression of the electrostatic self-potential energy of the particle, if the electric charge is distributed over a sphere of radius r is:

$$E_p(r) = \frac{e^2}{2r} = \frac{\alpha \hbar c}{2r} \tag{1a}$$

α is the fine structure constant:

$$\alpha = \frac{1}{137.04}$$

Assuming that the mass energy of the particle is of electrostatic origin, we can define the electrostatic energy of the field: $E_f(r)$, as the difference between the electrostatic energy and the mass energy of the particle.

$$E_f(r) = \frac{\alpha \hbar c}{2r} - mc^2 \tag{1b}$$

The basic principle of the model is to consider the quantum operator: $\hat{E}_f(r)$ associated with the energy defined above and to assume that we can define a momentum operator of the electrostatic field by the following relation:

$$\hat{A}c = \pm 4\pi\theta \hat{E}_f(r) \tag{2}$$

This equation is the same type as that which connects the momentum vector and the energy of the photons in the framework of the quantization of the free electromagnetic field (see below).

The parameter θ characterizes the particle and will be defined later for each particle.

We then pass to the equation of wave in spherical coordinates:

$$i\hbar c \frac{d(r\psi)}{rdr} = \pm 4\pi\theta E_f(r)\psi \tag{3}$$

Solutions to this differential equation appear immediately:

$$r\psi(r) < \exp[\pm \frac{4\pi i\theta}{\hbar c} \int_{r_0}^{r} E_f(r)dr] \tag{4}$$

We see below that the lower integration range is very small according to the constraint due to the model, of the order of magnitude of a Planck length.

Considering the real wave function, the sum of the function written above and its complex conjugate, we obtain:

$$r\psi(r) < \cos[\frac{4\pi\theta}{\hbar c} \int_{r_0}^{r} E_f(r)dr] \tag{5}$$

The relation (3) shows that the derivative of the above function is zero when the field potential energy defined in (1b) is zero, and therefore for the following value of the distance r to the particle:

$$r_m = \frac{\alpha \hbar}{2mc} \tag{6}$$

For this value of r, called "classical radius of particle", the electrostatic energy of the particle equals its mass energy, but it has not physical sense in classical physics.

The derivative of a function is null at its extrema, so the above value corresponds to an extremum. Since the function defined in (5) is a cosine function, this occurs when its argument is a multiple of π and for the value of r given above, which defines a quantization relation:

$$\frac{4\pi\theta}{hc} \int_{r_0}^{r_m} E_f(r)dr = n\pi \ ; \ n = integer \tag{7}$$

Using (6) and (7), it becomes:

$$\ln\frac{r_m}{r_0} = \frac{na}{2\theta} + 1 \text{ with } a = \frac{1}{\alpha} = 137.04 \tag{8}$$

It is interesting to transform (8) by introducing the particle mass according to relation (6), the Planck mass and the Planck radius:

$$r_p = \sqrt{\frac{G\hbar}{c^3}} \quad m_p c r_p = \hbar$$

We obtain thus

$$\ln\frac{m_p}{m} = \frac{na}{2\theta} + 1 + \ln(2a) - \ln\frac{r_p}{r_0} \tag{9}$$

$$\frac{m}{m_p} = \exp\left(-\frac{na}{2\theta} + K\right) \tag{10}$$

$$K = -1 - \ln(2a) + \ln\frac{r_p}{r_0}$$

Therefore, the masses of charged leptons are functions of the elementary electrical charge.

Two parameters determine the mass of each of them: the quantum number n, deriving from the characteristics of the wave function and the parameter θ which will be determined here in after. The ratio between the Planck mass and the electron mass is:

$$\ln\frac{m_p}{m_1} = 51.5 \tag{11}$$

This value is obtained with equation (9) by giving the parameters the following values:

$$n = 3; \; \theta = 3\sqrt{2}; \; \frac{r_p}{r_o} = 35.5 \qquad (12)$$

The last datum means that the value of r corresponding to the lower integration range is less than Planck's radius; it shows, according to this model, that the core of the particle has a dimension of the order of magnitude of the Planck radius. However, the numerical value given here is only indicative because RCs are omitted and the metric around the Planck radius is not Euclidean, which changes the calculation of the integral in relation (7).

The ratio between the mass of each of the two heavy leptons (of indices j = 2,3) and that of the electron is obtained from equation (9):

$$\ln \frac{m_j}{m_1} = \frac{a}{2}\left(\frac{n_1}{\theta_1} - \frac{n_j}{\theta_j}\right); \; a = 137.04 \qquad (13)$$

This relationship will make it possible to calculate the mass of the muon and tau leptons, but it will be necessary to take into account the radiative corrections (RC) to Coulomb's law as described in the appendix.

3. Assumed Structure of the Elementary Particles and Determination of the Coefficients θ

3.1. Fundamentals

We published an article in *Physics Essays* [3] showing that elementary particles can be considered self-gravitating photons on a Planck scale. The model constructed from this hypothesis makes it possible to calculate the value of the elementary electrical charge. More specifically, the model describes particles as composed of three pairs of photons with circular trajectories around a common center of gravity. Each pair carries a third of the elementary electrical charge and therefore each photon of a pair, called "fundamental element", one sixth of this charge.

Thus, this approach justifies the linear relation initially introduced between the momentum and energy operators in the electrostatic field.

Many authors have studied the spherical shells of self-gravitating photons within the framework of general relativity, but without involving the quantum theory, which is necessary to calculate the numerical value of the elementary electrical charge. The latest of the articles concerned is cited in reference [4].

3.2. Representation of Particles

A positive fundamental element is represented by O, a negative by Θ
The charged leptons in the form of three pairs are shown below as follows:

Θ Θ Charged leptons: O O Charged anti-leptons:
Θ Θ 3 negative pairs   O O 3 positive pairs
Θ Θ   O O

The neutrinos permit two representations corresponding to two quantum states:

State 1 Θ O   State 2   O O
         Θ O   Θ O
         Θ O   Θ Θ
3 neutral pairs   1 neutral pair + 2 pairs of opposite charges

In order to pursue the reasoning, let us define the momentum operator corresponding to the sum of the 6 basic elements, unstructured, as follows:

$$\hat{A}_o c = 4\pi \hat{E}_f(r) \tag{14a}$$

The operator defined above in (2) for the particle in question is such that:

$$\hat{A} = \theta \hat{A}_o \tag{14b}$$

The parameter θ translates the way the fundamental elements are structured for a given particle, from the point of view of the operator $\hat{A}$.

4. STRUCTURE AND MASSES OF CHARGED LEPTONS

We have determined empirically that for charged leptons, the expression of θ depends upon accounting for the three pairs of photons, according to the following coefficients: for a pair: $\sqrt{2}$; for a quadruplet of two pairs: $2 = \sqrt{4}$; for the three pairs together: $\sqrt{6}$

4.1. Electron

⊖⊖ For the electron, the parameter θ is assumed to be the sum of the values corresponding

⊖⊖ to three separate pairs: $\theta_1 = 3\sqrt{2}$ (15)

⊖⊖ The quantum number n = 3 gives the numerical result presented above.

4.2. Tau

⊖⊖ For the tau, the parameter θ is assumed to be the sum of the value corresponding to a

Θ Θ   pair + that corresponding to the quadruplet of the other two grouped pairs: $\theta_3 = 2 + \sqrt{2}$

Θ Θ   The value n = 2 of the quantum number makes it possible to obtain the mass of the tau as a function of that of the electron with relation   (16)

4.3. Muon

Θ Θ   For the muon, the parameter θ is assumed to be the sum of the corresponding values

Θ Θ   to two quadruplets of 2 pairs, subtracting the value corresponding to the common pair

Θ Θ   to the two quadruplets, assumed to be equal to: $\sqrt{6}/3$. We obtain thus:

$$\theta_2 = \sqrt{4} + \sqrt{4} - \frac{\sqrt{6}}{3} = 4 - \frac{\sqrt{6}}{3} \tag{17}$$

The value n = 2 of the quantum number makes it possible to obtain the mass of the muon with relation (13).

The method is empirical but coherent because there are only three ways presented here to combine the three pairs of photons, individually and / or by quadruplet of two pairs. This justifies the existence of three, and only three, families of particles.

The calculation obtained from the relation (13) and taking into account the radiative corrections gives values very close to the experimental measurements:

Calculated mass of the muon: 105.9 MeV _ mass measured: 105.7 MeV

Calculated mass of the tau: 1781 MeV _ mass measured: 1777 MeV

5. STRUCTURE AND MASSES OF NEUTRINOS

Neutrinos are devoid of electrical charge. However, each fundamental element as defined previously has an electric charge, positive or negative, equal in absolute value to $|e|/6$. Its electrostatic energy is therefore equal, with respect to the potential energy defined in (1), to $E_p(r)/36$. Reasoning thus, the potential energy of all these six elements would be: $E_p(r)/6$. Moreover, if into defining the momentum operator all six permutations of the pairs are taken into account, we arrive at $E_p(r)$.

We will study this hypothesis, which means using the model of charged leptons for neutrinos. The calculation of the masses of neutrinos will therefore be done by means of a relation identical to relation (13):

$$\ln \frac{m_{vj}}{m_e} = \frac{a}{2}\left(\frac{n_e}{\theta_e} - \frac{n_{vj}}{\theta_{vj}}\right); j = 1,2,3\,; a = 137.04 \tag{18}$$

Let us add that the radiative corrections will not intervene in this calculation because they concern only the charged leptons whose mass is greater than that of the electron.

For the value of a quadruplet, we will apply an empirical rule: $\sqrt{v_1^2 + v_2^2}$ v_1, v_2: *values of pairs*.

5.1. Electron Neutrino

As for the electron, each pair is taken into account separately.

- ΘO   The value of a neutral pair is found empirically equal to 1, which gives: $\theta_{v1} = 3$
- ΘO   By applying the relation (18) with n = 3, one obtains the *mass of the electron neutrino:*

- ΘO $m_{v1} = 0.983$ meV $\hfill (19)$

5.2. Muon Neutrino

In the case of the muon, only the quadruplets of 2 pairs are taken into account to calculate θ. The same principle will be applied for its neutrino by considering two ways in the state 2 above to build the quadruplets.

```
a-    Ө Ө                    a -   Ө Ө
a -   Ө O - b                a -   O O - b
      O O - b                      O Ө - b
```

The value of quadruplets a and b is $\sqrt{2+1} = \sqrt{3}$
The value of the quadruplet a is $\sqrt{2+2} = 2$
The value of the quadruplet b is $\sqrt{3}$
The value of θ is obtained by adding the 4 values above and subtracting the value of a common pair, assuming to be neutral.

$$\theta_{v2} = 3\sqrt{3} + 1 \qquad (20)$$

With n = 6 for the quantum number, the *mass of the muon neutrino* calculated with (18) is:

$$m_{v2} = 8.60 \; meV \qquad (21)$$

$$m_{v2}^2 - m_{v1}^2 = 7.3 \; x \; 10^{-5} eV^2$$

The masses of neutrinos are unknown, only the differences between the squares of their masses can be measured, or rather between the masses defined in the framework of the "neutrino mixing paradigm" [6]
The experimental measurements give: $\Delta m_{21}^2 = 7.5 \pm 0.3 \; 10^{-5} eV^2$.

5.3. Tau Neutrino

It turns out that in the case of the neutrino-tau, the calculation of θ is based on the sum of 2 cases of the state 1 above. Case1: similar to that of the tau. Case2: similar to that of the muon.

Case 1	Case 2	
Θ O	Θ O	In case 1, the value of the quadruplet of 2 pairs is $\sqrt{2}$ and 1 for the pair: $\theta_1 = \sqrt{2} + 1$. In case 2, the value of a pair is subtracted from the sum of the 2 quadruplets $\theta_2 = 2\sqrt{2} - 1$
Θ O	Θ O	
Θ O	Θ O	

The value of θ results from the sum of these two quantum states:

$$\theta_{\nu 3} = 3\sqrt{2} \tag{22}$$

n = 4 will be the value retained for the quantum number

Relation (18) gives:

$$m_{\nu 3} = 49.5 \; meV \tag{23}$$
$$m_{\nu 3}^2 - m_{\nu 1}^2 = 2.45 \times 10^{-3} eV^2$$

The experimental measurements give in the framework of the "neutrino mixing paradigm" give:

$$\Delta m_{31}^2 = 2.5 \pm 0.1 \times 10^{-3} \; eV^2$$

The model gives for the sum of the masses of the 3 neutrinos: 59 meV. This value is in accordance with experimental data.

6. STRUCTURE AND MASSES OF QUARKS

The quarks carry an electrical charge: |e|/3 or 2|e|/3 and a color charge. In our model based on self-gravitating photon pairs, we interpret this fact in the following way. It is known that in quantum theory one cannot refer to the conventional notions of electrical vector and magnetic vector for the photon. However, the notions of "electric photon" and "magnetic photon" are distinguished, which suggests that the charge of color is related to the magnetic component of the photons. Moreover, the experimental data indicate that the numerical value of the coupling of the strong interaction tends to the constant α to the high energies (property of asymptotic freedom). We will translate these remarks as follows for the quarks:

- When the electric charge of the quark is |e|/3, a pair of fundamental elements is electric, the other 2 pairs are colored
- When the electric charge of the quark is 2|e|/3, two pairs are electric, the third is colored.
- The numerical value of the square of the color charge of a pair is equal to that of the square of the electrical charge.
- Thus, the model developed for leptons can also be applied to quarks, which means that the calculation of the mass of a quark will be done by applying a relation of the same type as relation (13):

$$\ln \frac{m_q}{m_e} = \frac{a}{2}\left(\frac{n_e}{\theta_e} - \frac{n_q}{\theta_q}\right) \qquad (24)$$

For the value of a colored quadruplet, we apply an empirical rule: $\sqrt{v_1 + v_2}$ v_1, v_2: $values\ of\ pairs$

For a mixed quadruplet, the rule is

$$\sqrt{v_1 + v_2^2}\ ; v_1: colored\ pair;\ v_2: electrical\ pair$$

The problem posed by the radiative corrections will be discussed below. The quarks can thus be represented as follows, the color charge being represented by a triangle:

```
O O                      Θ Θ
O O                      Δ Δ
Δ Δ                      Δ Δ
```

Quarks: up, charm, top: +2e/3   Quarks: down, strange, bottom: - e/3

6.1. Masses of Quarks in the Electron Family

To obtain the parameter θ of the up and down quarks, we will proceed as for the electron, that is, we will sum the coefficients of the three pairs, considered individually, assuming that the coefficient of a colored pair is equal to 3. We neglect the radiative corrections because very weak here:

Down

For the quark d: $\theta_d = 6 + \sqrt{2}$;   n = 5

The calculation with relation (24) gives: m_d = 4.81 MeV   (25)

The QCD calculation carried out by Christine Davies [5] gives for its part: m_d = 4.79 ± 0.16 MeV and the PDG an interval: 4.3 _ 5.2 MeV

Up

For the quark u; $\theta_u = 3 + 2\sqrt{2}$ n = 4
The calculation with relation (24) gives: $m_u = 2.13$ MeV   (26)

The calculation of QCD gives for its part: $m_u = 2.01 \pm 0.14$ MeV and the PDG an interval:

1.8 _ 2.8 MeV

6.2. Masses of Quarks in the Tau Family

In the case of the tau lepton, the calculation of θ is carried out by adding the coefficient (value) of a pair and that of a quadruplet of two pairs. The same principle is applied to calculate the masses of bottom and top quarks.

Bottom

This quark comprises according to our model 2 colored pairs and a negative electrical pair with coefficients below:

- Δ Δ The mixed quadruplet value is assumed to be $\sqrt{3+2} = \sqrt{5}$ and the colored pair 3
- Θ Θ Hence $\theta_b = 3 + \sqrt{5}$; n = 3 will be retained for the quantum number
- Δ Δ The calculation carried out with the values of these parameters gives $m_b = 5020$ *MeV*

The calculation of the radiative corrections (RCs) presented in the appendix concerns only the charged leptons. We must assume that the value of ln (RCs) is 1/3 that of the fictitious particle of electric charge e having the same mass as the bottom:

This would give for the mass: $m_b = 4770$ *MeV* which is very close to half the mass of the upsilon, particle which is a bottomonium, of mass $m_U = 9460$ *MeV*. The PDG gives the interval:

4150 _ 4220 *MeV*

Top

This quark has 2 positive electric pairs and 1 colored pair. The two diagrams below show two ways of combining the coefficients of the quadruplets and the pairs:

(a)   (b)
O O   O O   In a) the quadruplet value is: $\sqrt{3+2} = \sqrt{5}$ and
                the pair value $\sqrt{2}$
Δ Δ   O O   In b) the quadruplet value is: $\sqrt{2+2} = 2$ and the
                pair value 3
O O   Δ Δ θ is assumed to result from the sum of these 2 cases,
                with a factor 2 taking into account the possible
                configurations, namely: $\theta_t = 2(5 + \sqrt{5} + \sqrt{2})$;
                n = 9 will be the value retained for the quantum
                number.

The calculation by means of relation (24) gives $m_t = 185\ GeV$

The calculation of RCs is here also hypothetical; by making a hypothesis similar to that of the previous case, the result obtained for the mass of the top would be: $m_t = 175\ GeV$.

The precise experimental measure is: $m_t = 173.3\ GeV$.

6.3. Masses of Quarks in the Muon Family

The parameter θ of the muon is obtained by adding the coefficients of 2 quadruplets and subtracting that of the intersection pair. The same process will be used for the 2 quarks.

Strange

a-Θ Θ   The quadruplet value a) is $\sqrt{3+2} = \sqrt{5}$; for the b)
            $\sqrt{3+3} = \sqrt{6}$; for the pair 3
a-Δ Δ-b   It proves empirically that it is necessary to double the
            coefficient of the sum of both
    Δ Δ-b   blocks before subtracting that of the common pair.
            Thus $\theta_s = 2(\sqrt{5} + \sqrt{6}) - 3$; the quantum number taken into
            account: n = 4

The calculation by means of relation (24) gives: $m_s = 116.6$ MeV

With RCs, the mass should rather be: $m_s = 110$ MeV

For the PDG, the strange mass is situated in the interval: 92 _ 104 MeV

Charm

a-O O   The quadruplet value a) is: $\sqrt{2+2} = 2$ and for the b): $\sqrt{3+2} = \sqrt{5}$; the common pair

a-O O-b   value *is* $\sqrt{2}$. Proceeding as above: $\theta_c = 2(\sqrt{5}+1) - \sqrt{2}$

Δ Δ-b   The calculation with relation (24) gives: $m_c = 1260$ MeV with the quantum number n = 3

By subtracting RCs, the mass of the *charm* would be rather: $m_c = 1200$ MeV

For the PDG, the mass of the *charm* is situated in the interval: *1250 _ 1310 MeV*

7. SUMMARY TABLE

a) First Family

Particle	Parameter θ	Quantum number n	θ/n	Model mass	Experimental mass
Electron	$3\sqrt{2}$	3	1.414	Reference	0.511 MeV
Neutrino e	3	3	1	0.983 meV	< 60 meV
Up-quark	$3+2\sqrt{2}$	4	1.457	2.13 MeV	2.01 MeV (Davies)
Down-quark	$6+\sqrt{2}$	5	1.483	4.81 MeV	4.79 MeV (Davies)

b) All Elementary Fermions

Particle	Parameter θ	Quantum number n	θ/n	Model mass	Experimental mass
Electron	$3\sqrt{2}$	3	1.414	Reference	0.511 MeV
Muon	$4 - \dfrac{\sqrt{6}}{3}$	2	1.592	105.9 MeV	105.7 MeV
Tau	$2 + \sqrt{2}$	2	1.707	1781 MeV	1777 MeV
Neutrino e	3	3	1	0.983 meV	< 60 meV
Neutrino μ	$3\sqrt{3}+1$	6	1.0327	8.60 meV	< 60 meV
Neutrino τ	$3\sqrt{2}$	4	1.0607	49.5 meV	< 60 meV
Up	$3 + 2\sqrt{2}$	4	1.457	2.13 MeV	1.8 _ 2.8 MeV
Down	$6 + \sqrt{2}$	5	1.483	4.81 MeV	4.3 _ 5.2 MeV
Strange	$2(\sqrt{5} + \sqrt{6}) - 3$	4	1.593	110 MeV	92 _ 104 MeV
Charm	$2(\sqrt{5} + 1) - \sqrt{2}$	3	1.686	1200 MeV	1250 _ 1310 MeV
Bottom	$3 + \sqrt{5}$	3	1.745	4770 MeV	4150 _ 4220 MeV
Top	$2(5 + \sqrt{5} + \sqrt{2})$	9	1.922	175 GeV	173.3 GeV

8. STERILE NEUTRINOS

A number of anomalies concerning the radiation emitted by nuclear reactors prompted specialists to consider the existence of sterile neutrinos, which are heavier than the three known neutrinos and have no interaction other than gravity with ordinary matter.

Sterile neutrinos have also been considered to explain unusual X-rays from some galaxies. These heavy neutrinos could be candidates for dark matter.

This article does not attempt to solve this mystery. It considers only a few possible cases of sterile neutrinos in the framework of this model, whose masses calculated with this model correspond to the experimental data. One can imagine three states of sterile neutrinos based on colored pairs. Δ denotes a fundamental element of a color and $\bar{\Delta}$ its anti-color:

State 1	State 2	State 3
Δ Δ (R)	Δ X̄	Δ Δ
Δ Δ (G)	Δ X̄	Δ X̄
Δ Δ (B)	Δ X̄	X̄ X̄
3 different color pairs = neutral	3 neutral pairs	1 color pair + 1 anti-color pair + 1 neutral pair

As for the quarks, a pair value Δ Δ is assumed to be 3; a pair value Δ X̄ is assumed to be 2. A quadruplet value of two colored pairs is assumed to be $\sqrt{3+3} = \sqrt{6}$.

8.1. Galactic Sterile Neutrino

Let us make a calculation analogous to that of the muon. In state 1, add the coefficients of 2 quadruplets and subtract that of a pair, that is:

$$\theta_1 = 2\sqrt{3+3} - 3 = 2\sqrt{6} - 3$$

Let's do the same in state 2:
$$\theta_2 = 2\sqrt{2+2} - 2 = 4 - 2 = 2$$

By cumulating the 2 cases, θ is the sum of the 2 values above:

$$\theta_s = 2\sqrt{6} - 1$$

The calculation using the model gives, with n = 3, a mass:

$$m_{sc} = 7.14 \: keV$$

Let us consider the value given by the cosmic X-ray energy measurements (3.57 keV) considered to be the result of the decay of this hypothetical neutrino [7].

8.2. Sterile Neutrinos of Nuclear Reactors

Consider in state 3 the sum of the quadruplet (colored pair + anti-color pair) and the neutral pair. The parameter θ of the particle will be: $\theta_s = 2 + \sqrt{6}$

- In the case that n = 3, calculation of the mass by means of relation (24) gives:

$$m_{sb} = 4.824 \ MeV$$

This could be the mass of a heavy sterile neutrino corresponding to the 5 MeV bump measured in the detectors: 4.9 MeV [8].

- In the case that n = 4, the calculation gives with the same value of θ above:

$$m_{sl} = 0.997 \ eV$$

This corresponds to the mass evaluated at 0.965 eV on a detector, if it is attributed to a hypothetical light sterile neutrino [6]. Following this hypothesis, it should be noted that the detectors installed in the nuclear reactors give several values > 1 eV for the mass of a light sterile neutrino. This leads to the assumption that there are at least two neutrinos of this type, one towards 1 eV and the other having a mass greater than 1 eV. The values of the intermediate masses measured would then correspond to mixtures of these two neutrinos produced by the reactors.

Two sterile neutrinos with mass > 1 eV could be produced by the model.

- One corresponding to the state 1 described above, for which the three pairs are considered separately. Then $\theta = 3 \times 3 = 9$. With n =

8, the calculation gives a mass close to 2 eV for this sterile neutrino.
- The other having the same structure as the cosmic sterile neutrino, but constructed from states 2 and 3. We then obtain $\theta = 2\sqrt{5}$ and with n = 4 a mass of 1.36 eV.

8.3. Hypothetical Sterile Neutrinos - Summary Table

Type of neutrino	Parameter θ	Quantum number n	θ/n	Model mass	Exp. Mass
Galactic	$2\sqrt{6} - 1$	3	1.2997	7.14 keV	7.14 keV
5 MeV bump	$\sqrt{6} + 2$	3	1.483	4.824 MeV	4.9 MeV
Light A	$\sqrt{6} + 2$	4	1.1124	0.997 eV	0.965 eV
Light B	$2\sqrt{5}$	4	1.118	1.36 eV	1.32 eV
Light C	9	8	1.125	1.99 eV	/

9. COLD DARK MATTER (CDM) PARTICLES?

Some recent articles deduct of cosmic-ray measurements (positrons, electrons) that can exist (no evidence) very massive particles of dark matter.

a) Man Ho Chan [9] relives an "abrupt change in the spectral radio signal of cosmic rays". By mean of a model, he deducts that this change is perhaps caused by the annihilation of a particle of mass 25 GeV. The present model, using the state 1 above of sterile neutrinos, finds a particle with:

$\theta = 3 + \sqrt{3+3} = 3 + \sqrt{6}$
1 pair+1 quadruplet
The calculated mass with n = 3 by the model is 23.37 GeV.

b) A direct analysis of cosmic rays by mean of the Chinese satellite DAMPE [11] detects a break in the energetic spectrum, close to 0.9 TeV; H.E.S.S collaboration [12](detector in Namibia) also finds a break, but close to 0.6 TeV.

The present model, using the state 2 of sterile neutrinos (3 separate pairs) finds a particle with $\theta = 3 \times 3 = 6$ and $n = 3$; the calculated mass is 0.745 TeV.

c) DAMPE collaboration detected also a bump at 1.4 TeV.

The present model, considering the sum of states 1 and 2 above of sterile neutrinos, and for each state the sum of 2 cases (3 separate pairs _ 1 pair+1 quadruplet) gives:

$\theta = 3x3 + (3 + \sqrt{6}) + 3x2 + (2 + \sqrt{4}) = 22 + \sqrt{6}$

With $n = 12$, we obtain a mass of: 1.4 TeV.

Hypothetical CDM - Summary Table

Type of neutrino	Parameter θ	Quantum number n	θ/n	Model mass	Cosmic-ray data
CDMa	$3 + \sqrt{6}$	3	1.816	23.37 GeV	25 GeV
CDMb	6	3	2	0.745 TeV	0.6_0.9 TeV
CDMc	$22+\sqrt{6}$	12	2.037	1.4 TeV	1.4 TeV

Conclusion

The model presented is somewhat beyond the standard model; however it cannot be termed "new physics" since it appeals only to the simplest and best known concepts of quantum theory. It is limited to extending the notion of quantum momentum to a static field.

As we have pointed out, the determination of the parameters θ, calculating the masses of the elementary fermions as a function of the mass of the electron, by means of a single simple relation, is the result of empirical investigation. All the calculations use 4 empirically assigned values for the following pairs: electrical pair ($\sqrt{2}$); neutral electrical pair (1); colored pair (3); neutral colored pair (2). Rules for calculating values of quadruplets from pair values are also empirical. Quantum numbers are not free parameters because they must give observable particle masses

The consistency of the method is guaranteed by the following considerations:

- The coefficient of a pair or of a quadruplet is always the same when it appears in the parameters of different particles.
- For each family of particles, the calculation mode of θ is homogeneous: three separate pairs for the electron family; two quadruplets - one pair for the muon family; one pair + one quadruplet for the tau family.
- The results obtained by the calculation are very close to the experimental values.

However, a more thorough theoretical exploration will be necessary to provide a better foundation this model. The study of the states of polarization of self-gravitating photons at the base of the model is probably an important research path to pursue in this field.

APPENDIX: RADIATIVE CORRECTIONS TO COULOMB'S LAW

Electrically charged virtual particles are the cause of a quantum vacuum polarization modifying the effective value of an electric charge according to a function of the distance to this charge. We have taken this effect into account, called "radiative corrections to Coulomb's law" in our article cited in [1] for electron-positron virtual pairs.

This effect is expressed in the model by the following expression of the energy of the electric field created by the particle.

$$E_f(r) = \frac{\alpha \hbar c}{2r}\left[1 - \lambda ln\left(\frac{\gamma m_e c r}{\hbar}\right)\right] - mc^2$$

$$\lambda = \frac{2\alpha}{3\pi} m_e c r < \hbar \gamma = \exp(C + \frac{5}{6})$$

C: Euler's constant = 0.577.

Continuing the calculation as before, we obtain for the mass of the muon:

$$\ln\frac{m_\mu}{m_e} = \frac{a}{2}\left(\frac{n_e}{\theta_e} - \frac{n_\mu}{\theta_\mu}\right) - \frac{\lambda}{2}\ln^2\left(\frac{2am_\mu}{\gamma m_e}\right)$$

To calculate the mass of the tau, it is necessary to take into account the radiative corrections due to the virtual pairs of the muons-anti-muons, namely:

$$\ln\frac{m_\tau}{m_e} = \frac{a}{2}\left(\frac{n_e}{\theta_e} - \frac{n_\tau}{\theta_\tau}\right) - \frac{\lambda}{2}\left[\ln^2\left(\frac{2am_\tau}{\gamma m_e}\right) + \ln^2\left(\frac{2am_\tau}{\gamma m_\mu}\right)\right]$$

For the mass of the muon, we obtain: 105.9 MeV (for 105.7 MeV measured)

And for the mass of tau: 1781 MeV (for 1777 MeV measured)

The value of radiative corrections is important in the case of muon and tau. In their absence, we would have found a mass 7% higher for the muon and 16% higher for the tau. On the other hand, there is no RCs for neutrinos.

Concerning the quarks, RCs must be taken into account since these particles comprise a fractional electrical charge. However, in this case the screen effects due to the strong interaction also intervene, opposites to RCs generated by the charged leptons. These effects are due to complex QCD calculations that we have not carried out, limiting ourselves to gross estimates of global RCs.

REFERENCES

[1] Raymond Fèvre; A Model of the Masses of Charged Leptons; *Physics Essays;* Vol 27 N° 4 (December 2014) p 608_611.

[2] Raymond Fèvre; The Higgs Boson and the Signal at 750 GV, Composite Particles? *JFAP* Vol 3, N° 1 (2016).

[3] Raymond Fèvre; Photons Self-gravitating and Elementary Charge; *Physics Essays;* Vol 26 N° 1; P 3_6 (March 2013).

[4] Hakan Andrasson, David Fajam, Maximilian Thaller; *Models for Self-gravitating Photon Shells and Geons; Annales Henri Poincaré*, Feb. 2017.

[5] Christine Davies; *Standard Model Heavy Flavor Physics on the Lattice;* ArXiv: 1203.3862v1 (17 March 2012).

[6] Carlo Giunti; Theory and Phenomenology of Massive Neutrinos; *Conference of Ciudad de Mexico;* 30 January_3 February 2017.

[7] Kevork N. Abazajian; *Resonantly Produced 7 keV Sterile Neutrino Dark Matter Models and the properties of Milky Way Satellites;* ArXiv: 1403.0954v3.

[8] Patrick Huber; *Neos Data and the Origin of the 5 MeV Bump in the Reactor Antineutrino Spectrum;* ArXiv 1609.03910v2; 19 Jan 2017.

[9] *A Possible Signature of Annihilating Dark Matter Man Ho Chan;* arXiv: 1711.O4398v1; 15 November 2017.

[10] Direct Detection of a Break in the Teraelectronvolt Cosmic-Ray Spectrum of Electrons and Positrons; DAMP Collaboration; *Nature* 552, 63_66; 07 December 2017.

[11] The Energy Spectrum of Cosmic-Ray Electrons at TeV Energies; H.E.S.S. Collaboration; F. Aharonian et al.; *Phys.Rev.Letters* 101:261 104, 2008.

In: Leptons
Editor: Christopher M. Villegas
ISBN: 978-1-53614-929-6
© 2019 Nova Science Publishers, Inc.

Chapter 2

THE ORIGIN OF SU (5) SYMMETRY BY LEPTON ELECTRIC CHARGE SWAP QUATERNIONS ALGEBRA

E. Koorambas[*]

Computational Applications Group,
Division of Applied Technologies,
National Center for Science and Research 'Demokritos',
Aghia Paraskevi-Athens, Greece

ABSTRACT

In the present chapter, taking the SU(2) group of weak interactions in the presence of Electric Charge Swap (ECS)-symmetry as a starting point, we show that ordinary and non-regular (ECS) leptons are related by the ECS-rotational SO(3) group. By considering the ECS-Hamiltonian quaternions for leptons, we find that the SU(5) Grand Unified Theory (GUT) originates from the image of normalized quaternions group $N(Q_8)$

---

[*] E-mail: elias.koor@gmail.com.

under the ECS-rotations. Furthermore, SU(5) symmetry as well as its Standard Model (SM) subgroup are not fundamental symmetries, since they can be derived by ECS-Lepontic quaternions. This means that gluons and photons are not fundamental particles of nature.

Keywords: group theory, hypothetical particles, grand unified theories

1. INTRODUCTION

Theories of unification based on the simple symmetric group are known as *Grand Unified Theories* (GUT). In the year 1974 H. Georgi and Glashow [1] proposed a theory of unification based on the SU(5) symmetric group that contains the Standard Model (SM) group as a subgroup. It is this theory unifies all interactions except gravity, because at energy above 10^{15} GeV it the SU(5) symmetric group has one gauge coupling, while at this energy scale it breaks down spontaneously to the symmetric group of the SM [1, 2]. The SU(5) symmetric group is the smallest group which contains the SM gauge group; as such, it has the greatest predictive power. However, there are three problems with SU(5) theory: the neutrino now appears to have mass [2]; the predicted decay rate of the proton is much higher than the current observed limit [2]; there is no explanation of the three generations of fermions [2; for details on GUT see [3-18].

Electric Charged Swap (ECS) symmetry for the case of leptons has been proposed by the author [19]. ECS-transformation between ordinary families of leptons produces heavy neutral non-regular leptons of mass of order O(1TeV). These particles may form cold dark matter [19]. Furthermore, ECS symmetry explains certain properties of lepton families within the framework of superstring theories [20-24].

A-Wollmann Kleinert and F. Bulnes considered the Higgs mechanism to re-combine gauge fields of *SU* (2) and *U* (1), gauge groups through three classes of bosons, *W*, *Z*, and *A* [25]. Based on ECS symmetry (in this case of leptons [19]), they proposed leptons as the subtle Fermions [25].

Recently, a quark (q) and an ECS-quark (q˜)-bound state (qq˜) have been proposed by the author [26]. This suggestion explains the electrically charged charmonium Z^+_c (3, 9) meson as a charm quark (c) and charm ECS-quark (c˜)-bound state (cc˜). It also predicts that J/ψ and π^+ mesons are the decay products of a Z^+_c (3, 9), as it has been observed recently at BES III [52-54]. Furthermore, this suggestion predicts two new mesons: an electrically charged charmed $D^{-*+(zm)}$ meson and a neutral charmed $D^{*0(zm)}$ meson. A new kind of space-time curvature effects, caused by electric charge swap (ECS) transformations between families of leptons has also been investigated [48].

Quaternions, introduced by Hamilton in 1843 [44, 45], are a number system that extends the complex numbers. P.R. Girard [46] shows how various physical covariance groups - SO(3), the Lorentz group, the general relativity group, the Clifford algebra SU(2) and the conformal group - can be readily related to the quaternion group in modern algebra. The same author demonstrates how Einstein's equations of general relativity could be formulated within a Clifford algebra that is directly linked to quaternions [47].

In the present paper, taking as our starting point the SU(2) group of weak interactions in the presence of ECS-symmetry, we show that ordinary and non-regular (ECS) leptons are related by the ECS-rotational SO(3) group and explore some of the implications of this finding.

2. Fundamental of The Electric Charge Swap (ECS) Symmetry in Six-Dimensional Space-Time

We begin with the simplest set-up, where only the third family of leptons exists in the four-dimensional part of six-dimensional space-time [19].

Following Gogberashvilli et al. [27], we consider a six-dimensional spacetime with signature $(+,-,-,-,-,-)$. Einstein's equations in this spacetime have the form:

$$R_{AB} - \frac{1}{2}g_{AB}R = \frac{1}{M^4}(g_{AB}\Lambda + T_{AB}) \tag{1}$$

where M the six-dimensional fundamental scale, Λ is the cosmological constant and A,B are capital indices equal 0,1,2,3,4,5,

To split the six-dimensional space-time into four-dimensional and two-dimensional parts, we use the metric ansatz [27]:

$$ds^2 = \phi^2(\theta)g_{\mu\nu}(x^a)dx^\mu dx^\nu - \varepsilon^2(d\theta^2 + b^2\sin^2\theta d\varphi^2), \tag{2}$$

where ε and b are constants and $\phi(\theta)$ is the warp factor. This warp factor equals one at brane location $(\theta = 0)$ and decreases to zero in the asymptotic region $(\theta = \pi)$, at the south pole of the extra two-dimensional sphere. Here the metric of the ordinary four-dimensional $g_{\mu\nu}(x^a)$ has signature $(+,-,-,-)$, with $\alpha, \mu, \nu = 0,1,2,3$. The extra compact 2-manifold is parameterized by the spherical angles θ, ϕ ($0 \leq \theta \leq \pi, 0 \leq \phi \leq 2\pi$). This 2-surface is attached to the brane at point $\theta = 0$. When θ changes from 0 to π, therefore, the geodesic distance into the extra dimensions shifts from the north to the south pole of the 2-spheroid. For $b = 1$ in equation (2), the extra 2-surface is exactly a 2-sphere with radius ε (0.1TeV^{-1}).

The ansatz for the energy-momentum tensor of the bulk matter fields is:

$$T_{\mu\nu} = -g_{\mu\nu}E(\theta), \quad T_{ij} = -g_{ij}P(\theta), \quad T_{i\mu} = 0. \tag{3}$$

Small latin indices in equation (3) correspond to the two extra coordinates. The source functions E and P depend only on the extra coordinate θ. For these ansätze, Einstein's equations (1) take the following form:

$$3\frac{\phi''}{\phi} + 3\frac{\phi'^2}{\phi^2} - 3\frac{\phi'}{\phi}\cot\theta - 1 = \frac{\varepsilon^2}{M^4}[E(\theta) - \Lambda],$$

$$6\frac{\phi'^2}{\phi^2} - 4\frac{\phi'}{\phi}\cot\theta = \frac{\varepsilon^2}{M^4}[P(\theta) - \Lambda],$$

$$4\frac{\phi''}{\phi} + 6\frac{\phi'^2}{\phi^2} = \frac{\varepsilon^2}{M^4}[P(\theta) - \Lambda]. \qquad (4)$$

where the prime denotes differentiation d/dθ.

For the four-dimensional space-time, we have assumed zero cosmological constant. Einstein's equations take the form:

$$R^{(4)}_{\alpha\beta} - \frac{1}{2}g_{\alpha\beta}R^{(4)} = 0, \qquad (5)$$

where $R^{(4)}_{\alpha\beta}$ and $R^{(4)}$ are four-dimensional Ricci tensor and scalar curvature, respectively. In [28] M. Gogberashvili and D. Singleton found a non-singular solution of (4) for boundary conditions $\phi(0) = 1$, $\phi'(0) = 0$. This solution was given by:

$$\phi(\theta) = 1 + (a-1)\sin^2(\theta/2), \qquad (6)$$

where a is the integration constant. The source terms for this solution were given by:

$$E(\theta) = \Lambda\left[\frac{3(a+1)}{5\phi(\theta)} - \frac{3a}{10\phi^2(\theta)}\right], \quad P(\theta) = \Lambda\left[\frac{4(a+1)}{5\phi(\theta)} - \frac{3a}{5\phi^2(\theta)}\right], \qquad (7)$$

with the radius of the extra 2-spheroid given by $\varepsilon^2 = 10M^4/\Lambda$.

For simplicity, in this paper we take $a = 0$ so that the warp factor takes the form:

$$\phi(\theta) = 1 - \sin^2(\theta/2) = \cos^2(\theta/2). \tag{8}$$

This warp factor equals one at the brane location ($\theta = 0$) and decreases to zero in the asymptotic region $\theta = \pi$, i.e., at the south pole of the extra two-dimensional spheroid. The expression for the determinant of the ansatz (2) used in this paper is given by:

$$\sqrt{-g} = \sqrt{-g^{(4)}}\, \varepsilon^2 \phi^4(\theta) \sin\theta, \tag{9}$$

where $\sqrt{-g^{(4)}}$ is the determinant of four-dimensional space-time.

2.1. Non-Regular Leptons in Six Dimensions

Here we assume that the zero mode corresponds to the non-regular leptons which are copies of the third family of leptons. Although arbitrary, this assumption is not physically implausible: it is reasonable to expect that upon entering the six-dimensional bulk, third family leptons change their properties profoundly and lose, so to speak, their individuality (e.g., their observable masses); they are reduced to their bare mass, spin and magnetic moment [19].

Let us now consider spinors in the six-dimensional space-time (2), where the warp factor $\phi(\theta)$ has the form (8). The action integral for the six-dimensional massless fermions in a curved background is:

$$S_\psi = \int d^6 x \sqrt{-g} \left[i\bar{\Psi} h_{\tilde{A}}^B \Gamma^{\tilde{A}} D_B \Psi + h.c \right] \tag{10}$$

D_A is the covariant derivative and $\Gamma^{\tilde{A}}$ is the 6-dimensional flat gamma matrices. We have also introduced the sechsbein $h_A^{\tilde{A}}$ through the usual definition [27].

$$g_{AB} = h_A^{\tilde{A}} h_B^{\tilde{B}} \eta_{\tilde{A}\tilde{B}}, \tag{11}$$

where $\tilde{A}, \tilde{B}$ are local Lorenz index.

The six-dimensional spinor is given by:

$$\Psi(x^A) = \begin{pmatrix} \psi \\ \xi \end{pmatrix}. \tag{12}$$

This six-dimensional spinor has eight components and is equivalent to a pair of four-dimensional Dirac spinors, ψ, ξ. The representation of the flat (8 × 8) gamma-matrices is given by [27] as:

$$\Gamma_v = \begin{pmatrix} \gamma_v & 0 \\ 0 & -\gamma_v \end{pmatrix}, \Gamma_\theta = \begin{pmatrix} 0 & -1 \\ 1 & 0 \end{pmatrix}, \Gamma_\theta = \begin{pmatrix} 0 & -i \\ i & 0 \end{pmatrix}, \tag{13}$$

where 1 denotes the four-dimensional unit matrix and γ_v are ordinary (4×4) gamma-matrices. Representation (13) gives the correct space-time signature $(+,-,-,-,-,-)$. The generalization of γ_5 matrix is:

$$\Gamma_7 = \begin{pmatrix} \gamma_5 & 0 \\ 0 & \gamma_5 \end{pmatrix}. \tag{14}$$

The variation of action (10) yields the following six-dimensional massless Dirac equation:

$$(h_{\tilde{B}}^\mu \Gamma^{\tilde{B}} D_\mu + h_{\tilde{B}}^\theta \Gamma^{\tilde{B}} D_\theta + h_{\tilde{B}}^\varphi \Gamma^{\tilde{B}} D_\varphi) \Psi(x^A) = 0, \tag{15}$$

with the sechsbein for our background metric (2) given by

$$h_{\tilde{A}}^{B} = \left(\frac{1}{\phi}\delta_{\tilde{\mu}}^{B}, \frac{1}{\varepsilon}\delta_{\tilde{\theta}}^{B}, \frac{1}{\varepsilon\sin\theta}\delta_{\tilde{\phi}}^{B}\right) \tag{16}$$

From the definition of spin connection:

$$\omega_{M}^{\tilde{M}\tilde{N}} = \frac{1}{2}h^{N\tilde{M}}(\partial_{M}h_{N}^{\tilde{N}} - \partial_{N}h_{M}^{\tilde{N}}) - \frac{1}{2}h^{N\tilde{N}}(\partial_{M}h_{N}^{\tilde{M}} - \partial_{N}h_{M}^{\tilde{M}})$$
$$-\frac{1}{2}h^{P\tilde{M}}h^{Q\tilde{N}}(\partial_{P}h_{Q\tilde{R}} - \partial_{Q}h_{P\tilde{R}})h_{M}^{\tilde{R}}. \tag{17}$$

The non-vanishing components of the spin connection are:

$$\omega_{\varphi}^{\tilde{\theta}\tilde{\varphi}} = -\sin\theta \;,\; \omega_{\nu}^{\tilde{\theta}\tilde{\nu}} = -\frac{\varphi'}{\varepsilon} = \frac{\sin\theta}{2\varepsilon} \tag{18}$$

The covariant derivatives of the spinor field take the form:

$$D_{\mu}\Psi(x^{A}) = (\partial_{\mu} + \frac{\sin\theta}{4\varepsilon}\Gamma_{\theta}\Gamma_{\nu})\Psi(x^{A})$$
$$D_{\theta}\Psi(x^{A}) = \partial_{\theta}\Psi(x^{A}) \tag{19}$$
$$D_{\varphi}\Psi(x^{A}) = (\partial_{\varphi} - \frac{\cos\theta}{2}\Gamma_{\theta}\Gamma_{\varphi})\Psi(x^{A})$$

According to [29] and [30], the Dirac equation takes the form:

$$\left(\frac{1}{\varphi}\Gamma^{\mu}\frac{\partial}{\partial x_{\mu}} + \frac{\sin\theta}{4\varepsilon\varphi}\Gamma^{\mu}\Gamma_{\theta}\Gamma_{\varphi} + \frac{1}{\varepsilon}\Gamma^{\theta}\frac{\partial}{\partial\theta} + \frac{1}{\varepsilon\sin\theta}\Gamma^{\varphi}\frac{\partial}{\partial\varphi} - \frac{\cot\theta}{2\varepsilon}\Gamma^{\varphi}\Gamma_{\theta}\Gamma_{\varphi}\right)\Psi(x^{A})$$
$$= \left[\frac{1}{\varphi}\Gamma^{\mu}\frac{\partial}{\partial x_{\mu}} + \frac{1}{\varepsilon}\Gamma^{\theta}\left(\frac{\partial}{\partial\theta} - \frac{\sin\theta}{\varphi} + \frac{\cot\theta}{2}\right) + \frac{1}{\varepsilon\sin\theta}\Gamma^{\varphi}\frac{\partial}{\partial\varphi}\right]\Psi(x^{A}). \tag{20}$$

This system of first-order partial differential equations has the following solutions:

$$\Psi(x^A) = \frac{1}{\sqrt{2\pi\phi^2(\theta)}} \begin{pmatrix} \alpha_0(\theta)\psi_0(x^\nu) \\ \beta_0(\theta)\xi_0(x^\nu) \end{pmatrix}, \tag{21}$$

with $\psi_0(x^\nu), \xi_0(x^\nu)$ being the four-dimensional Dirac spinors.

We note that since the dimensions of $\Psi(x^A)$ in six dimensions are $m^{5/2}$, the dimensions of $\alpha_0(\theta), \beta_0(\theta)$ and $\psi_0(x^\nu), \xi_0(x^\nu)$ should be m and $m^{3/2}$, respectively.

We are looking for four-dimensional leptonic zero modes. To this end, we consider the conditions under which equation (21) obeys the four-dimensional, massless Dirac equations;

$$\gamma^\mu \partial_\mu \psi_0(x^\nu) = \gamma^\mu \partial_\mu \xi_0(x^\nu) = 0 \tag{22}$$

Of course, there are also very massive Kaluza Klein (KK) modes of masses n/ε. However, since we assume that $1/\varepsilon \approx 10 \text{TeV}$, these massive KK modes have a much higher mass and are distinct from the third family of leptons.

For the massless case, the 4 spinors $\psi_0(x^\nu), \xi_0(x^\nu)$ are indistinguishable from the four-dimensional point of view, and we can write $\psi_0(x^\nu) = \xi_0(x^\nu)$. Inserting (21) and (22) into (20) converts the bulk Dirac equation into:

$$\left[\Gamma^\theta \left(\frac{\partial}{\partial \theta} \right) + \cot\theta \right] \begin{pmatrix} \alpha_0(\theta) \\ \beta_0(\theta) \end{pmatrix} = 0 \tag{23}$$

Using the representation for Γ^θ, Γ^ϕ gives the following system of equations for $\alpha_0(\theta)$ and $\beta_0(\theta)$

$$\left(\frac{\partial}{\partial \theta}-\frac{\cot\theta}{2}\right)\alpha_0(\theta)=0, \quad \left(\frac{\partial}{\partial \theta}+\frac{\cot\theta}{2}\right)\beta_0(\theta)=0 \qquad (24)$$

The solutions of these equations are:

$$\alpha_0(\theta)=\frac{A_0}{\sqrt{\sin\theta}}, \quad \beta_0(\theta)=\frac{B_0}{\sqrt{\sin\theta}}, \qquad (25)$$

where A_0 and B_0 are integration constants with the dimension of mass. The normalizable modes are those for which:

$$\int \sqrt{-g}\, d^6 x\, \bar{\Psi}\Psi = \int \sqrt{g^{(4)}}\, d^4 x\left(\bar{\psi}_0\psi_0+\bar{\xi}_0\xi_0\right) \qquad (26)$$

In other words, we want the integral over the extra coordinates, φ and θ to equal 1. Inserting (21), (25) and the determinant (9) into (26), the latter requirement gives:

$$\pi\varepsilon^2(A_0^* A_0 + B_0^* B_0)=1 \qquad (27)$$

Explicitly, the expressions for the three normalizable 8-spinors (21) that solve the six-dimensional Dirac equations (20) are:

$$\Psi_0(x^A)=\frac{1}{\sqrt{2\pi\sin\theta\phi^2(\theta)}}\binom{A_0}{B_0}\psi_0(x^\nu) \qquad (28)$$

where constants A_0 and B_0 obey the relations (27).

2.2. The Set-Up of the ECS Symmetry in Six-Dimensional Space-Time

In the four-dimensional part of six-dimensional space-time, non-regular leptons have the same mass as ordinary third family leptons. Hypothetical non-regular leptons are, a) a zero-charged version of the tau, $\tilde{\tau}^0$ (1784MeV) and, b) a positively charged version of the tau neutrino, $\tilde{v}_\tau^+$ (0,1eV). Non-regular leptons can, therefore, be obtained from the swap of electric charges between tau and tau neutrino particles in the six-dimensional part. We call these proposed non-regular leptons, electric charge swap (ECS) leptons [19].

Although ECS leptons have the same mass as ordinary third family leptons, they are distinguished from the latter by their different lepton numbers ($L_s = 1$ for ordinary leptons and $\bar{L}_s = -1$ for ordinary antileptons, respectively) and by their electric charges (positive or neutral for ordinary leptons; negative or neutral for ordinary antileptons, respectively). We hypothesize that ECS leptons are produced from third family leptons when these enter the six-dimensional bulk: in these conditions, the properties of third family leptons change profoundly as these leptons lose, so to say, their individuality and swap their electric charge [19].

To formulate the swap of electric charge between ordinary leptons, we have to look for symmetry characteristic of the swap process in the framework of 2-extra dimensions with compactification scale 10 TeV [19].

We consider the 2-sphere S^2 as a quotient space $S^2 \equiv SU(2)_L / U(1)_Y$ and express this space in terms of the symmetry between the original lepton and the new, ECS lepton doublets. To do this, we proceed through the following steps [19]:

Firstly, we observe that both the ordinary lepton doublet $l_0(x^v) = (\tau_L^-, v_\tau)$ and the ECS lepton doublet $\tilde{l}_0(x^{v'}) = (\tilde{\tau}_L^0, \tilde{v}_\tau^+)$ can form the fundamental representation of $SU(2)_L$ [31].

This fundamental representation is given by:

$$[I_j, I_k] = i\varepsilon_{jkl} I_l .\tag{29}$$

The generators are denoted as:

$$I_i = \frac{1}{2}\tau_i .\tag{30}$$

where

$$\tau_1 = \begin{pmatrix} 1 & 0 \\ 0 & 1 \end{pmatrix}, \tau_2 = \begin{pmatrix} 0 & -i \\ i & 0 \end{pmatrix}, \tau_3 = \begin{pmatrix} 1 & 0 \\ 0 & -1 \end{pmatrix}\tag{31}$$

are the isospin versions of Pauli matrices.

The action of the latter on the new leptons states is represented by:

$$\tilde{\tau}_L^0 = \begin{pmatrix} 1 \\ 0 \end{pmatrix}, \tilde{v}_\tau^+ = \begin{pmatrix} 0 \\ 1 \end{pmatrix}\tag{32}$$

To link the two distinct sectors, ordinary and ECS leptons, we assume that neither ordinary L nor ECS L_s lepton numbers are conserved, while the overall lepton number is conserved obligatorily.

$$L_{overall} = L_s + L = 0 .\tag{33}$$

$$L_s = \bar{L}, L_s(\tilde{v}_\tau^+) = \bar{L}(\tau^+) = -1\tag{34}$$

$$\bar{L}_s = L, \bar{L}_s(\tilde{\tau}^0) = L(v_\tau) = 1 .\tag{35}$$

The quantum numbers of the new ECS leptons of mass 1784 MeV and 0,1eV respectively, are given in Table 1 [19].

The next step is to define the group transformation that can account for the swap of electric charges between the tau and tau neutrino particles. The ECS transformation must be derived from a transformation from

1. $SU(2)_L / U(1)_Y$, in which the fundamental representation of $SU(2)_L$ is $l_0(x^v) = (\tau_L^-, v_\tau)$ and $U(1)_Y$ is the symmetric group generated by hypercharge $Y = -1$ to,

2. $SU(2)_L / U(1)_{Y_S}$, in which the fundamental representation of $SU(2)_L$ is $\tilde{l}_0(x^{v'}) = (\tilde{\tau}_L^0, \tilde{v}_\tau^+)$ and $U(1)_{Y_S}$ is the symmetric group generated by swap hypercharge $Y_S = 1$.

The quotient space $SU(2)/U(1)$ is diffeomorphic to the unit 2-sphere S^2. Consequently, the swap of the electric charges between the tau and neutrino of tau particles must be an automorphism of the 2-sphere to itself [19].

Table 1. Quantum numbers (mass M, weak isospin I, charge Q, hypercharge Ys, Lepton number Ls) of the ECS leptons $\tilde{\tau}_L^0, \tilde{v}_\tau^+$

New lepton	M	I	I-z	Q	Ys	Ls
$\tilde{v}_\tau^+$	0,1eV	½	½	1	1	-1
$\tilde{\tau}_L^0$	1784MeV	½	-½	0	1	-1

Since the two extra dimensions are endowed with the Fubini-Study[1] metric [32, 33], not all Möbius transformations (e.g., dilations and

---

[1] The round metric of the 2-extra dimensional sphere can be expressed in stereographic coordinates as $G = \dfrac{dy_1^2 + dy_2^2}{(1+\varepsilon^2)^2}$, where $\varepsilon = \sqrt{y_1^2 + y_2^2}$. The metric G is the Fubini-Study metric of the 2-sphere [32, 33].

translations) are isometries. Therefore, the automorphism from the $S^2 \equiv SU(2)/U(1)$ to itself, which brings the electric charge swap between the tau and neutrino of tau particles, is given by the isometries that form a proper subgroup of the group of projective linear transformations $PGL_2(\hat{C})_{(Charge)}$, namely $PSU_{2(Charge)}$ Subgroup $PSU_{2(Charge)}$ is isomorphic to the rotation group $SO(3)^{(ECS)}$ [32, 33,19], which is the isometric group of the unit sphere in three-dimensional real space $\mathbb{R}^3$. The automophism of the Riemann sphere $\hat{C}$ is given by:

$$Rot_{(ECS)}(\hat{C}) = PSU_{2(Charge)} = SO(3)^{(ECS)}, \qquad (36)$$

$$\hat{C} = \mathbb{C} \cup \infty = S^2$$

where $\hat{C}$ is the extended complex plane, $PSU_{2(Charge)}$ is the proper subgroup of the projective linear transformations and swap symmetry, $SO(3)^{(ECS)}$ is the group of rotations in three-dimensional vector space $\mathbb{R}^3$. The universal cover of $SO(3)^{(ECS)}$ is the special unitary group $SU(2)^{(ECS)}$ [19]. This group is also differomorphic to the unit 3-sphere S^3.

---

where $SU(2)^{(ECS)}$ is the special unitary group, and $U(1)_{Y_3(Y_{3s})}$ is the symmetric group generated by hypercharge $Y_3(Y_{3s})$.

ECS symmetry has been tested in various numbers of lepton families and space-time dimensions [19]. This symmetry explains some properties of lepton families within the framework of superstring theories [20-24]. Furthermore, a mechanism has been proposed to explain local symmetry breaking at energy scales below 14TeV [19]. Local symmetry breaking can make the (postulated) non-regular leptons highly massive, 'explaining' thereby why these leptons are unobservable by Large Electron Positron ring (LEP) I, II and neutrino oscillations experiments at energy scales below 14TeV [34-37, 38-40]. The existence of these proposed leptons can be tested once the Large Hadron Collider (LHC) becomes operative at 14 TeV energy-scales [41-43].

We regard ordinary and ECS leptons as different electric charge states of the same particle – an analogy with the proton-neutron isotopic pair. Finally, in terms of rotational symmetry between the original lepton and the proposed ECS leptons, the ECS two-extra dimensional sphere S^2_{3-ECS} is given by:

$$S^2_{3-ECS} \equiv SU(2)^{(ECS)} / U(1)_{Y_3(Y_{3s})} [19] \tag{37}$$

3. THE ECS-HAMILTONIAN QUATERNION

The basis of ECS gauge group O (3) $_{ECS}$ is given by:

$$L = \begin{pmatrix} \tilde{v}_e^+ \\ v_e \sin\theta_s + \tilde{e}^0 \cos\theta_s \\ e^- \end{pmatrix}_{Left}, R = \begin{pmatrix} \tilde{v}_e^+ \\ \tilde{e}^0 \\ e^- \end{pmatrix}_{Right},$$

$$\left(\tilde{e}^0 \sin\theta_s - v_e \cos\theta_s\right)_{Left}, \tag{38}$$

where θ_s is the swap angle, L, R are the left and right triplet respectively and $\left(\tilde{e}^0 \sin\theta_s - v_e \cos\theta_s\right)_{Left}$ is the left singlet.

Let the ECS-Hamiltonian quaternion be defined as follows:

$$H_{ECS} := \mathbb{R}1 \oplus \mathbb{R}(\tilde{v}_e^+)i \oplus \mathbb{R}(\tilde{e}^0)j \oplus \mathbb{R}(e^-)k , \tag{39}$$

where

$$H = \mathbb{R}1 \oplus \mathbb{R}i \oplus \mathbb{R}j \oplus \mathbb{R}k , \tag{40}$$

and

$$(1, R^T) = (1, \tilde{v}_e^+, \tilde{e}^0, e^-) \tag{41}$$

H is the quaternion algebra in a 4-dimensional real vector space with basis $1, i, j, k$ [49, 50]; $(1, R^T)$ is a basis that contains the ECS-leptons $\tilde{v}_e^+, \tilde{e}^0$ and an electron, e^-.

The multiplication rule of the H quaternion algebra is [50]:

$$ij = k, jk = i, ki = j, i^2 = j^2 = k^2 = -1. \tag{42}$$

Equation (42) induces the following multiplication rule for the ECS-quaternion:

$$(i\tilde{v}_e^+)(j\tilde{e}^0) = e^- k, (j\tilde{e}^0)(ke^-) = \tilde{v}_e^+ i, (ke^-)(i\tilde{v}_e^+) = j\tilde{e}^0, \tag{43}$$

$$(i\tilde{v}_e^+)^2 = (j\tilde{e}^0)^2 = (ke^-)^2 = -1, \tag{44}$$

This multiplication rule is extended to H_{ECS} via the associative and distributive law. The subalgebra $\mathbb{R} = \mathbb{R}1$ is the central of H_{ECS}. For unique $t, x, y, z \in \mathbb{R}$ every ECS-quaternion $q_{ECS} \in H_{ECS}$ may thus be expressed as:

$$q_{ECS} = t + x(\tilde{v}_e^+)i + y(\tilde{e}^0)j + z(e^-)k \tag{45}$$

The conjugate of (45) is the quaternion:

$$\overline{q}_{ECS} = t - x(\tilde{v}_e^-)i - y(\overline{\tilde{e}}^0)j - z(e^+)k. \tag{46}$$

The norm of q_{ECS} is:

$$N(q_{ECS}) = q_{ECS}\overline{q}_{ECS} \in \mathbb{R}. \tag{47}$$

Hence $0 \leq N(q_{ECS})$, with equality only for $q_{ECS} = 0$.

We also note that

$$N(p_{ECS} q_{ECS}) = N(p_{ECS}) N(q_{ECS}). \tag{48}$$

It follows that, if $q_{ECS} \neq 0$, then $N(q_{ECS})^{-1} \cdot \bar{q}_{ECS}$ is the multiplication inverse of q_{ECS} in H_{ECS}.

Hence H_{ECS} is the division algebra whose set of nonzero elements,

$$H_{ECS}^{\times} = H_{ECS} - \{0\}, \tag{49}$$

is a group under ECS-quaternion multiplication.

The norm N is homomorphism

$$N : H_{ECS}^{\times} \to \mathbb{R}_{>0}^{\times}, \tag{50}$$

from $H_{ECS}^{\times}$ to the group $\mathbb{R}_{>0}^{\times}$ of positive real numbers under multiplication, whose kernel is

$$\begin{aligned} KerN &= \{q_{ECS} \in H_{ECS}^{\times} \mid q_{ECS} \bar{q}_{ECS} = 1\} \\ &= \{t + x(\tilde{v}_e^+) i + y(\tilde{e}^0) j + z(e^-) k \in H_{ECS} \mid t^2 + x^2 + y^2 + z^2 = 1\}. \end{aligned} \tag{51}$$

4. THE LIE GROUP S_{ECS}^3

(Equation 51) can be identified with the 3-dimensional ECS-sphere $S_{ECS}^3 \subset \mathbb{R}^4$. The 3-dimensional ECS-sphere S_{ECS}^3 may, therefore, be written as follows:

$$S^3_{ECS} = \{q_{ECS} \in H^\times_{ECS} \mid q_{ECS}\bar{q}_{ECS} = 1\}. \tag{52}$$

S^3_{ECS} is thus a group under ECS-quaternion multiplication, fitting into the exact sequence:

$$1 \to S^3_{ECS} \to H^\times_{ECS} \xrightarrow{N} \mathbb{R}^\times_{>0} \to 1. \tag{53}$$

Group S^3_{ECS} contains as a subgroup the ECS-quaternion group:

$$Q^{ECS}_8 = \{\pm 1, \pm \tilde{v}^+_e i, \pm \tilde{e}^0 j, \pm \tilde{e}^- k\}, \tag{54}$$

of order eight. S^3_{ECS} is, therefore, nonabelian, and the central of S^3_{ECS} has just two elements:

$$Z(S^3_{ECS}) = \{\pm 1\}. \tag{55}$$

Since (55) is already the full central of Q^{ECS}_8, subgroup

$$T_{ECS-i} = \{t + x(\tilde{v}^+_e)i \mid t^2 + x^2 = 1\} = \{e^{i\theta} \mid \theta_s \in \mathbb{R}\}, \tag{56}$$

is abelian subgroup of S^3_{ECS}, isomorphic to S^1_{ECS}, the ECS-circle group with θ_s swap angle. Note that

$$T_{ECS-i} = C_{S^3_{ECS}}(\tilde{v}^+_e i) \tag{57}$$

is the centralizer of ECS-lepton ($\tilde{v}^+_e i$) in S^3_{ECS}.

Furthermore, we have the three subgroups

$$T_{ECS-i}, T_{ECS-j}, T_{ECS-k} \subset S^3_{ECS}, \tag{58}$$

all isomorphic to S^1_{ECS}.

$$\begin{aligned} T_{ECS-i} &= \{e^{i\theta_s} \mid \theta_s \in \mathbb{R}\}, \\ T_{ECS-j} &= \{e^{j\theta_s} \mid \theta_s \in \mathbb{R}\}, \\ T_{ECS-k} &= \{e^{k\theta_s} \mid \theta_s \in \mathbb{R}\}, \end{aligned} \tag{59}$$

where

$$e^{i\theta_s} = \cos\theta_s + i\sin\theta_s, \, e^{j\theta_s} = \cos\theta_s + j\sin\theta_s, \, e^{k\theta_s} = \cos\theta_s + k\sin\theta_s \tag{60}$$

5. The Normalizer of Q_8^{ECS}

Following [49, 50], $\tilde{v}_e^+ i, \tilde{e}^0 j, e^- k$, as well as subgroups $T_{ECS-i}, T_{ECS-j}, T_{ECS-k}$, are conjugate in S^3_{ECS}. On the other hand, the relation in the ECS-quaternion group Q_8^{ECS} suggests that Q_8^{ECS} has an ECS-automorphism of order three, sending

$$\tilde{v}_e^+ i \to \tilde{e}^0 j \to e^- k \to \tilde{v}_e^+ i \tag{61}$$

Is this ECS-automorphism realized by conjugation inside the normalizer $N(Q_8^{ECS})$ of Q_8^{ECS} in S^3_{ECS}? If so, then, since each circle T_{ECS*} is the centralizer of corresponding subscript *, the same element would conjugate

$$T_{\tilde{v}_e^+ i} \to T_{\tilde{e}^0 j} \to T_{e^- k} \to T_{\tilde{v}_e^+ i}. \tag{62}$$

There are exactly two elements $q_{ECS} \in S^3_{ECS}$ that satisfy

$$q_{ECS}\tilde{v}_e^+ i q_{ECS}^{-1} = \tilde{e}^0 j, \quad q_{ECS}\tilde{e}^0 j q_{ECS}^{-1} = e^- k, \quad q_{ECS} e^- k q_{ECS}^{-1} = \tilde{v}_e^+ i.; \quad (63)$$

namely $\pm\frac{1}{2}\left(1+\tilde{v}_e^+ i+\tilde{e}^0 j+e^- k\right)$, which have orders six (+) and three (-).

Elements $q_{ECS} = \pm\frac{1}{2}\left(1+\tilde{v}_e^+ i+\tilde{e}^0 j+e^- k\right) \in S^3_{ECS}$, with all possible combinations of sings, will conjugate $\tilde{v}_e^+ i, \tilde{e}^0 j, e^- k$ with each other, up to sign that makes element q_{ECS} a normalizer of Q_8^{ECS}, acting on Q_8^{ECS} via outer automorphism. These elements, along with Q_8^{ECS} itself, comprise the normalizer of Q_8^{ECS} Thus, $N(Q_8^{ECS})$ consists of 24 ECS-quaternions:

$$N(Q_8^{ECS}) = \left\{1 \pm \tilde{v}_e^+ i \pm \tilde{e}^0 j \pm e^- k\right\} \bigcup \left\{\frac{1}{2}\left(1 \pm \tilde{v}_e^+ i \pm \tilde{e}^0 j \pm e^- k\right)\right\}, \quad (64)$$

and Q_8^{ECS} is the Sylow 2-subgroup of $N(Q_8^{ECS})$. The group $N(Q_8^{ECS})$ is the binary ECS-tetrahedral group, given as follows:

$$N(Q_8^{ECS}) \equiv SL_2(\mathbb{Z}/3\mathbb{Z})_{ECS} \quad (65)$$

This is the group of 2×2 matrices over $\mathbb{Z}/3\mathbb{Z}$ with determinant equal to one.

6. THE ORIGIN OF SU (5) SYMMETRIC GROUP BY ECS- QUATERNION ALGEBRA

Let us regard $\mathbb{R}^3$ as the ECS-sphere of the pure quaternions:

$$H_0^{ECS} := \mathbb{R}(\tilde{v}_e^+)i \oplus \mathbb{R}(\tilde{e}^0)j \oplus \mathbb{R}(e^-)k = \{u \in H_{ECS} \mid \tau(u) = 0\} \quad (66)$$

where τ is the trace and u is a vector in H_0^{ECS}. The dot product between two vectors $u, v \in H_0^{ECS}$, may be expressed quaternionically as

$$\langle u, v \rangle = \frac{1}{2}(u\bar{v} + v\bar{u}) \, [50]. \quad (67)$$

For $q_{ECS} \in S_{ECS}^3$, let $R_q^{ECS} : H_0^{ECS} \to H_0^{ECS}$ be the linear mal given by

$$R_q^{ECS} = q_{ECS} u q_{ECS}^{-1} \quad (68)$$

Using (78), we have

$$\langle u, v \rangle = \langle R_q^{ECS}(u), R_q^{ECS}(v) \rangle \quad (69)$$

for all $u, v \in H_0^{ECS}$ and $q_{ECS} \in S_{ECS}^3$.

It is, therefore, $R_q^{ECS} \in O(3)_{ECS}$, and we have the continuous homomorphism

$$R_q^{ECS} : S_{ECS}^3 \to O(3)_{ECS}. \quad (70)$$

Sending $q_{ECS} \to R_q^{ECS}$, since S_{ECS}^3 is connected, the image of ECS-rotation R_q^{ECS} is also connected. R_q^{ECS} thus lies within the $SO(3)_{ECS}$ (the subsets $O(3)_{ECS}, SO(3)_{ECS} \subset M_n(\mathbb{R})$ are compact and $SO(3)_{ECS}$ is connected, while $O(3)_{ECS}$ has two connected components).

To demonstrate the homomorphism explicitly, let

$$q_{ECS} = au + b\tilde{v}_e^+ i + c\tilde{e}^0 j + de^- k \in S_{ECS}^3, \quad (71)$$

and

$$q_{ECS}\tilde{v}_e^+ i q_{ECS}^{-1} = (a^2 + b^2 - c^2 - d^2)\tilde{v}_e^+ i + 2(bc + ad)\tilde{e}^0 j + 2(bd - ac)e^- k, \quad (72)$$

$$q_{ECS}\tilde{e}^0 j q_{ECS}^{-1} = 2(bc - ac)\tilde{v}_e^+ i + (a^2 - b^2 + c^2 - d^2)\tilde{e}^0 j + 2(cd + ab)e^- k, \quad (73)$$

$$q_{ECS} e^- k q_{ECS}^{-1} = 2(bd + ac)\tilde{v}_e^+ i + 2(cb - ab)\tilde{e}^0 j + (a^2 - b^2 - c^2 - d^2)e^- k. \quad (74)$$

The matrix of the ECS-rotation R_q^{ECS} with respect to basis $\{\tilde{v}_e^+ i, \tilde{e}^0 j, e^- k\}$ is given by

$$R_q^{ECS} = \begin{pmatrix} a^2 + b^2 - c^2 - d^2 & 2(bc - ac) & 2(bd + ac) \\ 2(bc + ad) & a^2 - b^2 + c^2 - d^2 & 2(cb - ab) \\ 2(bd - ac) & 2(cd + ab) & a^2 - b^2 - c^2 - d^2 \end{pmatrix} \quad (75)$$

Let $S_{ECS}, T_{ECS} \in SO(3)_{ECS}$ be ECS-rotations by swap angles $2\theta_s, 2\phi_s$ about unit vectors $u, v \in H_0^{ECS}$. Then $S_{ECS} = R_p^{ECS}, T_{ECS} = R_q^{ECS}$ where

$$q_{ECS} = \cos\theta_s + u\sin\theta_s, \quad (76)$$

$$p_{ECS} = \cos\phi_s + u\sin\phi_s \quad (77)$$

From (76) ad (77), it follows that $(ST)_{ECS} = R_{pq}^{ECS}$.
We compute

$$(pq)_{ECS} = (\cos\theta_s \cos\phi_s) - (\sin\theta_s \sin\phi_s)u \cdot v + (\sin\theta_s \cos\phi_s)u$$
$$+ (\cos\theta_s \sin\phi_s)v + (\sin\theta_s \sin\phi_s)u \times v \qquad (78)$$

With rotation by swap-angle ψ_s about the axis through the vector w, we have

$$\cos\psi_s = \cos\theta_s \cos\phi_s - (\cos\theta_s \sin\phi_s)u \cdot v, \qquad (79)$$

and

$$w = (\sin\theta_s \cos\phi_s)u + (\cos\theta_s \sin\phi_s)v + (\sin\theta_s \sin\phi_s)u \times v \qquad (80)$$

The image under ECS-rotation R_{ECS} of a binary tetrahedral group $N(Q_8^{ECS})$ (Equation (65)) is a symmetric group of a regular tetrahedron and isomorphic to the ECS-alternative group A_4^{ECS}. Thus we have the sequence

$$1 \to \{\pm 1\} \to N(Q_8^{ECS}) \to A_4^{ECS} \to 1, \qquad (81)$$

where $A_4^{ECS} \sim SU(5)^{ECS}$ the ECS-alternative group A_4^{ECS}, isomorphic to the ECS-special unitary group $SU(5)^{ECS}$.

Sequence (81) is not split: there is not subgroup of $N(Q_8^{ECS})$ isomorphic to $A_4^{ECS} \sim SU(5)^{ECS}$. $N(Q_8^{ECS})$ and S_4^{ECS}, in particular, are non-isomorphic groups of order 24. Note that the latter group fits into another exact sequence, which splits

$$1 \to A_4^{ECS} \to S_4^{ECS} \to \{\pm 1\} \to 1 \qquad (82)$$

In $SU(5)^{ECS}$, the fundamental representation takes the form

$$\tilde{5}_F = \begin{bmatrix} d_c \\ \varepsilon_2 \tilde{L} \end{bmatrix}_R, \text{ where } \tilde{L} = \begin{pmatrix} \tilde{v}_e^+ \\ \tilde{e}^0 \end{pmatrix}. \tag{83}$$

Here, the subscripts c refer to colors red, blue, green; $\tilde{L}$ is the ECS-lepton doublet.

Table 2. Quantum numbers that form the fundamental representation of the 24 gauge bosons

$SU(5)^{ECS}$	d_r	d_b	d_g	$\tilde{v}_e^+$	$\tilde{e}^0$
d_r	$g, \gamma \tilde{Z}$	g	g	X_r	Y_r
d_b	g	$g, \gamma \tilde{Z}$	g	X_b	Y_b
d_g	g	g	$g, \gamma \tilde{Z}$	X_g	Y_g
$\tilde{v}_e^+$	X_r	X_b	X_g	$\gamma, \tilde{Z}$	$\tilde{W}^+$
$\tilde{e}^0$	Y_r	Y_b	Y_g	$\tilde{W}^-$	$\gamma, \tilde{Z}$

The ECS-particle and the SM particle can be contained in the multiples of $SU(5)^{ECS}$:

$$\overline{\tilde{5}}_F = \begin{bmatrix} \overline{d}_c \\ \varepsilon_2 \tilde{l} \end{bmatrix}_L, \quad \tilde{10}_F = \begin{pmatrix} \varepsilon_3 \tilde{u}_c & q_c \\ -q^T_c & \varepsilon_2 \tilde{v}_e^- \end{pmatrix}. \tag{84}$$

This relationship is reproduced for each of the three SM families and their ECS-copies. $\tilde{10}_F$ is an antisymmetric multiple derived from

$$\tilde{5}_F \otimes \overline{\tilde{5}}_F = 1\tilde{5}_F \oplus \tilde{10}_F, \tag{85}$$

where $1\tilde{5}_F$ is totally symmetric. There are $5^2-1=24$ gauge bosons. In terms of their fundamental representation, they have the form shown in Table 2.

There are, therefore, eight gluons g; three weak ECS-bosons $\tilde{W}^-$, $\tilde{W}^+$, $\tilde{Z}$; and one photon γ, along with 12 X,Y-bosons. The ECS-SU(5) symmetric group is the smallest group which contains the ECS-SM gauge group [25]. Since we do not observe SU(5) ECS-symmetry, it means that the latter may be broken. ECS-SM symmetry breaking can occur at some high energy scale; electroweak ECS-symmetry breaking can also occur, at lower scale. The breaking of SU(5) symmetry down to electromagnetism is under investigation [51].

CONCLUSION

In the present chapter, taking the SU(2) group of weak interactions in the presence of ECS-symmetry as our starting point, we show that ordinary and non-regular (ECS) leptons are related by the ECS-rotational SO(3) group. By considering the ECS-Hamiltonian quaternions for leptons, we find that the SU(5) GUT symmetry originates from the image of normalized quaternions group $N(Q_8)$ under the ECS-rotations. Furthermore, the SU(5) symmetry as well as its SM subgroup are not fundamental symmetries, since they can be derived by ECS-leptonic quaternions. This means that gluons and photons are not fundamental particles of nature.

DEDICATION

This work is dedicated to Konstantina Marketou, and to our garden.

REFERENCES

[1] Georgi H., Glashow. S. L, *Phys. Rev. Lett.* 32 (1974) 451.
[2] Mohapatra R. N., Unification and supersymmetry. The frontiers of quark-leptons physics Springer - Verlag (1986), G.G. Ross, *Grand Unification Theories,* Benjamin/Cummings (1985).
[3] Lucas V. and Raby. S, "Nucleon Decay in a Realistic SO(10) SUSY Gut," *Phys. Rev.* D 55 (1997) 6986 [arXiv:hep-ph/9610293].
[4] Goto T. and Nihei.T, "Effect of Rrrr Dimension Five Operator on the Proton Decay in the Minimal SU(5) Sugra GUT Model," *Phys. Rev.* D59 (1999) 115009 [arXiv:hep-ph/9808255].
[5] Hisano J., Murayama. H and Yanagida. T, "Nucleon Decay in the MinimalSupersymmetric SU(5) Grand Unification," *Nucl. Phys.* B 402(1993) 46 [arXiv:hep-ph/9207279].
[6] Murayama H. and Pierce. A, "Not Even Decoupling Can Save Minimal Supersymmetric SU(5)," *Phys. Rev.* D 65 (2002) 055009 [arXiv:hepph/0108104].
[7] Bajc. B, Fileviez. P Perez. P and Senjanović. G, "*Minimal Supersymmetric SU(5) Theory and Proton Decay: Where Do We Stand?,*" arXiv:hepph/0210374.
[8] Marciano W. J. and Senjanović. G, "Predictions of Supersymmetric Grand Unified Theories," *Phys. Rev.* D 25 (1982) 3092.
[9] Weinberg S., "Does Gravitation Resolve the Ambiguity Among Supersymmetry Vacua?," *Phys. Rev. Lett.* 48 (1982) 1776-1779.
[10] Giudice G. F. and Rattazzi. R, "R-Parity Violation and Unification," *Phys. Lett.* B 406 (1997) 321 [arXiv:hep-ph/9704339].
[11] Aulakh. C. S and Mohapatra. R. N, "Implications of Supersymmetric SO(10) Grand Unification," *Phys. Rev.* D 28 (1983) 217.
[12] Clark T. E., Kuo. T. K. and Nakagawa. N, "A SO(10) Supersymmetric Grand Unified Theory," *Phys. Lett.* B 115 (1982) 26.
[13] Wilczek F. and Zee. A, "Families from Spinors," *Phys. Rev.* D 25 (1982) 553.

[14] Aulakh C. S. and Girdhar. A, "SO(10) a La Pati-Salam," *Int. J. Mod. Phys.* A 20 (2005) 865 [arXiv:hep-ph/0204097].

[15] Mohapatra R. N. and Senjanović. G, "Neutrino Mass and Spontaneous Parity Nonconservation," *Phys. Rev. Lett.* 44 (1980) 912.

[16] Mohapatra R. N. and Senjanović. G, "Neutrino Masses and Mixings in Gauge Models with Spontaneous Parity Violation," *Phys. Rev.* D 23 (1981) 165.

[17] Lazarides G., Shafi. Q and Wetterich. C, "Proton Lifetime and Fermion Masses in an SO(10) Model," *Nucl. Phys.* B 181 (1981) 287.

[18] Babu K. S. and R. Mohapatra. R. N, "Predictive Neutrino Spectrum in Minimal SO(10) Grand Unification," *Phys. Rev. Lett.* 70 (1993) 2845 [arXiv:hep-ph/9209215].

[19] Koorambas E, *Commun. Theor. Phys.* 60 (2013) 561–570

[20] Polykov A. M., *Phys. Lett.* B. 103 (1981).

[21] Szabo R. J., *arXiv*: hep-th/0207142 (2002).

[22] Giveon A., Kutasov. D, *Reviews of Modern Physics,* Vol. 71 No. 4 (1999).

[23] Polchinski J., *Phys. Rev. Lett.* 75 (1995) 4724.

[24] Polchinski J., *arXiv*: hep-th/9611050 (1996).

[25] Koorambas E. *International Journal of High Energy Physics. Special Issue: Symmetries in Relativity, Quantum Theory, and Unified Theories.* 2, No. 4, (2015). 1-7.

[26] Kleinert. A. W, Bulnes. F., *Journal on Photonics and Spintronics*, 2,1, (2013).

[27] Gogberashvilli M., Midodashvili. M, Singleton. D, *JHEP* 08 (2007) 033.

[28] Gogberashvili M. and Singleton. D, *Phys. Lett.* B 582 (2004) 95.

[29] Camporesi R. and Higuchi. A, *J. Geom. Phys.,* 20 (1996) 1.

[30] Abrikosov (jr) A. A., *Int. J. Mod. Phys.* A 17 (2002) 885.

[31] Halzen F, Martin. A. D, *Quarks and Leptons: An Introduction Course in Modern Particle Physics* (John Wiley & Son, New York USA 1984) 251-252,292-301.

[32] Fubini G., *Atti Istit.Veneto,* 63 (1904) 502.
[33] Study E., *Math. Ann.* 60 (1905) 321.
[34] Physics at LEP2, *CERN Yellow Report* 96–01 (2001).
[35] de Rujula A. et al., *Nucl. Phys.* B 384 (1992) 3.
[36] Hagiwar K. et al., *arXiv*: 0611102 [hep-ph] (2006).
[37] Particle Data Group, *J. Phys.* G 33 1 (2006) 1120-1147.
[38] Super-Kamiokande Collaboration (Ashie.Y et al.), *Phys. Rev.* D 71 (2005) 112005.
[39] Ambrosio M. et al. [MACRO Collaboration], *Eur. Phys. J.* C 36 (2004) 323.
[40] Sanchez M. C. et al. [Soudan 2 Collaboration], *Phys. Rev.* D 68 (2003) 113004.
[41] Aleksa M. et al., *Jinst*/P0403 (2008).
[42] Tarem S., Bressler. S, *ATL –SN – 071* (2008).
[43] Ellis J. *Phil. Trans. R. Soc.* A (2012) 370, 818-830, Virdee.T.S, *Phil. Trans. R. Soc.* A (2012) 370, 876-891.
[44] *On Quaternions; or on a new System of Imaginaries in Algebra* (letter to John T. Graves, dated 17 October 1843). 1843.
[45] Abramovich Rozenfel′d Boris (1988). *The history of non-euclidean geometry: Evolution of the concept of a geometric space.* Springer. p. 385. ISBN 9780387964584.
[46] Girard P. R., "The quaternion group and modern physics." *European Journal of Physics.* 5 (1984) 25–32.
[47] Girard, P. R. "Einstein's equations and Clifford algebra" (PDF). *Advances in Applied Clifford Algebras.* 9 (2) (1999) 225–230. Archived from the original (PDF) on 17 December 2010.
[48] Koorambas E., *Journal of Magnetohydrodynamics, Plasma, and Space Research* 20, 3, (2015) 333.
[49] Reeder M. *Math 845 Introduction to Lie Groups* (Course notes) Boston College (December 2010).
[50] Bar-Itzhack Itzhack Y. "New method for extracting the quaternion from a rotation matrix," *AIAA Journal of Guidance, Control and Dynamics,* 23 .6, (2000) 1085–1087.

[51] Koorambas E., *The doublet/Triplet splitting problem in the presence of quark and leptons of swap electric charge* (in preparation).

[52] Ablikim M. et al. [BESIII Collaboration]. *Phys. Rev. Lett.* 110 (2013) 252001, arXiv:1303.5949 [hep-ex], (2013).

[53] Choi S. K. et al. [BELLE Collaboration]. *Phys. Rev. Lett.* 100 (2008) 142001, [arXiv:0708.1790 [hep-ex].

[54] Mizuk R. et al. [Belle Collaboration], *Phys. Rev.* D 78 (2008) 072004, [arXiv:0806.4098hep-ex].

In: Leptons
Editor: Christopher M. Villegas

ISBN: 978-1-53614-929-6
© 2019 Nova Science Publishers, Inc.

Chapter 3

SCATTERING $\nu_{\mu,\tau} - e^-$ IN THE 331RHν MODEL AND ELECTROMAGNETIC PROPERTIES

A. Gutiérrez-Rodríguez[1,*] *A. Burnett-Aguilar*[1] *and M. A. Hernández-Ruíz*[2]

[1]Unidad Académica de Física,
Universidad Autónoma de Zacatecas,
Zacatecas, México
[2]Unidad Académica de Ciencias Químicas,
Universidad Autónoma de Zacatecas
Zacatecas, México

Abstract

We calculate the differential cross-section corresponding to the dispersion process $\nu_{\mu,\tau} - e^-$ in the context of the 331RHν model in order to use our results in terrestrial and astrophysical experiments. The differential cross-section is written in terms of the mass of the new gauge boson Z', the mixing angle ϕ, the magnetic moment of the neutrino μ_ν, and the charge radius $\langle r^2 \rangle$. Furthermore, our results are compared with those corresponding to the Standard Model in the decoupling limit $M_{Z'} \to \infty$ and $\phi = 0$.

---

*E-mail address: alexgu@fisica.uaz.edu.mx.

1. Introduction

The Neutrinos n the Standard Model (SM) [1, 2, 3] are massless particles which interact very weakly with matter via the exchange between the Z and $W^{\pm}$ bosons. Of all the different properties of the neutrino, its electromagnetic properties are the least known experimentally. In addition, the observation of neutrino oscillation shows the necessity of neutrino masses, which implies that the SM should be modified so that the non-trivial electromagnetic structure of neutrino may be reconsidered [4]. In the minimal extension of the SM, the incorporation of the neutrino mass allows the anomalous magnetic moment (MM) to be obtained with the loop calculation, $\mu_\nu = \frac{3eG_F m_{\nu_i}}{(8\sqrt{2}\pi^2)} \simeq 3.1 \times 10^{-19}(\frac{m_{\nu_i}}{1\,eV})\mu_B$, where $\mu_B = \frac{e}{2m_e}$ is the Bohr magneton [5, 6]. The non-zero mass of the neutrino is essential to get a non-vanishing magnetic moment. Furthermore, the SM predicts the CP violation which is necessary for the existence of the electric dipole moment (EDM) in a variety of physical systems. The EDM provides a direct experimental probe of CP violation [7, 8, 9] and is a feature of the SM and of physics beyond the SM. The signs of new physics can be analyzed by investigating the electromagnetic dipole moments of the neutrino such as the MM and the EDM. The study of the electromagnetic properties of the neutrino could help not only to shed light on whether neutrinos are Dirac or Majorana particles, but also to constrain the existing beyond Standard Model theories.

The most extensively applied procedure for the experimental investigation of neutrino electromagnetic properties involves direct laboratory measurements of the low energy elastic scattering of neutrinos and antineutrinos with electrons in reactor, accelerator, and solar experiments. These properties can also be studied in the astrophysical system where neutrinos propagate in strong magnetic fields and in dense matter [10]. The present work aims to point out the effects of neutrino electromagnetic interactions on the differential cross-section of the scattering experiments via the process

$$\nu_l + e^- \to \nu_{l'} + e^-, \qquad (1)$$

in the context of the 331RHν model, where a neutrino or antineutrino with flavor $\nu_l = \nu_\mu, \nu_\tau$ and energy E_ν elastically scatters off a free electron at rest in the laboratory frame.

The experimental evidence on the phenomenon of neutrino oscillation is a clear sign that the SM of elementary particle physics must be extended. Although there are many extensions of the SM, the models based on the gauge group $SU(3)_C \otimes SU(3)_L \otimes U(1)_X$ [11, 12, 13, 14], also known as 331 models, are one of the simplest extensions of the SM. These electroweak models have several interesting characteristics. For example, both the asymptotic freedom and the cancellation of the chiral anomalies require that the number of triplets of the fermion must be equal to the number of anti-triplets. Due to the incorporation of two different scales of energy, an explanation is thus given for the large difference between the SM particle masses of the third family and those of the other two families. Furthermore, since the model introduces a scalar sector similar to that of the two Higgs doublet model (2HDM), it is possible to predict the quantization of the electric charge as well as the vectorial character of the electromagnetic interaction without considering the nature of the neutrinos. There are different versions of the 331 model, namely $\beta = \pm\sqrt{3}, \pm 1/\sqrt{3}$. A common characteristic of these models is the fulfillment of a fundamental relation of the gauge group generators

$$Q = T^3 + \beta T^8 + X, \qquad (2)$$

where Q indicates the electric charge, T^3 and T^8 are two of the generators of the group $SU(3)$, X is the generator of the group $U(1)_X$ and β is the parameter that specifies the model under consideration. In this document, we consider the 331RHν model with a $\beta = 1/\sqrt{3}$ parameter.

Our main objective in this work is to provide suitable expressions for the differential cross-section of the process $\nu_{\mu,\tau} + e^- \to \nu_{\mu,\tau} + e^-$ in the context of a 331 model. These expressions can be easily incorporated into realistic neutrino experiments and give important information regarding electromagnetic properties.

The document is organized as follows: in Sec. 2, general formulas for the total differential cross-sections are presented and the conclusions are given in Sec. 3.

2. Neutrino-Electron Scattering

2.1. Vertex of Interaction $\nu\bar{\nu}\gamma$

In order to determine the effects on the anomalous magnetic and electric dipole moments of the ν, we calculate the differential cross-section of the reactions

$\nu_{\mu,\tau} + e^- \to \nu_{\mu,\tau} + e^-$. We use the electromagnetic current parametrization for the coupling $\nu\bar{\nu}\gamma$ of the photon to neutrino, given by

$$\begin{aligned}\Gamma^\alpha_\nu &= eF_1(q^2)\gamma^\alpha + \frac{ie}{2m_\nu}F_2(q^2)\sigma^{\alpha\mu}q_\mu + \frac{e}{2m_\nu}F_3(q^2)\sigma^{\alpha\mu}q_\mu\gamma_5 \\ &+ eF_4(q^2)\gamma_5(\gamma^\alpha - \frac{2q^\alpha m_\nu}{q^2}),\end{aligned} \quad (3)$$

where e is the charge of the electron, m_ν is the mass of the neutrino, q^μ is the photon momentum, and $F_{1,2,3,4}(q^2)$ are the electromagnetic form factors of the neutrino corresponding to charge radius, magnetic moment (MM), electric dipole moment (EDM) and anapole moment (AM), respectively, at $q^2 = 0$ [4, 15, 16, 17, 18, 19]. The form factors corresponding to the charge radius and the anapole moment are not considered in this work.

2.2. Differential Cross-Section of $\nu_{\mu,\tau} + e^- \to \nu_{\mu,\tau} + e^-$

For our study, we will take advantage of the previous works on the collision mode $\nu_l + e^- \to \nu_{l'} + e^-$ [14] to calculate the differential cross-section for the $\nu_{\mu,\tau} + e^- \to \nu_{\mu,\tau} + e^-$ reactions. The corresponding Feynman diagrams for this process are given in Fig. 1.

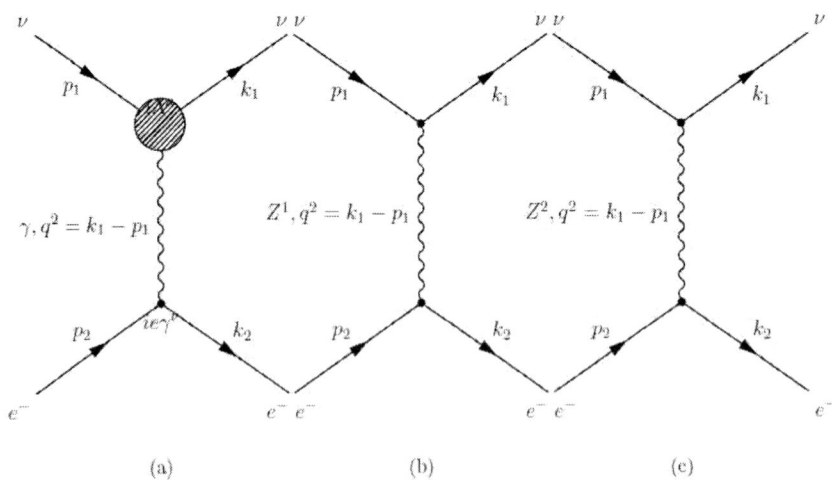

Figure 1. Feynman diagrams for the scattering $\nu_{\mu,\tau} + e^- \to \nu_{\mu,\tau} + e^-$.

The total amplitude of transition for the process $\nu_{\mu,\tau} + e^- \to \nu_{\mu,\tau} + e^-$ is given by

$$\mathcal{M}_{Total} = \mathcal{M}_\gamma + \mathcal{M}_Z + \mathcal{M}_{Z'}, \tag{4}$$

where we include the contribution of the electromagnetic interaction of the photon with the neutrinos Γ^α, as well as the contribution of the neutral currents (Z, Z') in the context of the model 331RHν. $\mathcal{M}_\gamma$, $\mathcal{M}_Z$ and $\mathcal{M}_{Z'}$ are explicitly given by

$$\mathcal{M}_\gamma = -\frac{e^2}{q^2}\bar{u}(k_1)\Lambda^\mu(l,q)u(p_1)\bar{u}(k_2)\gamma_\mu u(p_2), \tag{5}$$

$$\begin{aligned}\mathcal{M}_Z &= \sqrt{2}G_F[\bar{u}(k_1)\gamma^\mu(g^\nu_{1V} - g^\nu_{1A}\gamma_5)u(p_1)] \\ &\times [\bar{u}(k_2)\gamma_\mu(g^e_{1V} - g^e_{1A}\gamma_5)]u(p_2)],\end{aligned} \tag{6}$$

$$\begin{aligned}\mathcal{M}_{Z'} &= \sqrt{2}G_F\frac{M_Z}{M_{Z'}}[\bar{u}(k_1)\gamma^\mu(g^\nu_{2V} - g^\nu_{2A}\gamma_5)u(p_1)] \\ &\times [\bar{u}(k_2)\gamma_\mu(g^e_{2V} - g^e_{2A}\gamma_5)u(p_2)],\end{aligned} \tag{7}$$

where the new couplings of the Z, Z' bosons with the SM fermions are given in Table 1. The couplings $g^e_{(1V,2V)}$ ($g^\nu_{(1V,2V)}$) and $g^e_{(1A,2A)}$ ($g^\nu_{(1A,2A)}$) depend on the $Z - Z'$ mixing angle ϕ. When $\phi = 0$ and $M_{Z'} \to \infty$ in the decoupling limit, the SM couplings are recovered. The square of the total transition amplitude is

$$\begin{aligned}|\mathcal{M}|^2 &= \mathcal{M}\mathcal{M}^\dagger, \\ &= (\mathcal{M}_\gamma + \mathcal{M}_Z + \mathcal{M}_{Z'})(\mathcal{M}_\gamma + \mathcal{M}_Z + \mathcal{M}_{Z'})^\dagger, \\ &= |\mathcal{M}_\gamma|^2 + |\mathcal{M}_Z|^2 + |\mathcal{M}_{Z'}|^2 + |\mathcal{M}_{int}|^2.\end{aligned} \tag{8}$$

We then have the following for the first term of Eq. (8)

$$\begin{aligned}|\mathcal{M}_\gamma|^2 &= \frac{e^4}{q^4}[\bar{u}(k_1)\Gamma^\mu(l,q)u(p_1)\bar{u}(k_2)\gamma_\mu u(p_2)] \\ &\times [\bar{u}(p_2)\gamma_\nu u(k_2)\bar{u}(p_1)\Gamma^\nu(l,-q)u(k_1)].\end{aligned} \tag{9}$$

Because they are unobservable variables, we add and average over the particle spin states, obtaining

Table 1. The new couplings of the Z, Z' bosons with the SM fermions in the 331RHν model with ϕ as the $Z - Z'$ mixing angle [14]

Particle	Couplings
e^-, μ, τ	$g_{1V} = (-\frac{1}{2} + 2\sin^2\theta_W)(\cos\phi - \frac{\sin\phi}{\sqrt{3-4\sin^2\theta_W}})$, $g_{1A}^\nu = -\frac{1}{2}(\cos\phi - \frac{\sin\phi}{\sqrt{3-4\sin^2\theta_W}})$
ν_e, ν_μ, ν_τ	$g_{1V}^\nu = \frac{1}{2}(\cos\phi + \sin\phi\sqrt{3-4\sin^2\theta_W})$, $g_{1A}^\nu = \frac{1}{2}(\cos\phi + \sin\phi\sqrt{3-4\sin^2\theta_W})$
e^-, μ, τ	$g_{2V} = (-\frac{1}{2} + 2\sin^2\theta_W)(\sin\phi - \frac{\cos\phi}{\sqrt{3-4\sin^2\theta_W}})$, $g_{2A}^\nu = -\frac{1}{2}(\sin\phi - \frac{\cos\phi}{\sqrt{3-4\sin^2\theta_W}})$
ν_e, ν_μ, ν_τ	$g_{2V}^\nu = \frac{1}{2}(\sin\phi + \cos\phi\sqrt{3-4\sin^2\theta_W})$, $g_{2A}^\nu = \frac{1}{2}(\sin\phi + \cos\phi\sqrt{3-4\sin^2\theta_W})$

$$\begin{aligned}\overline{|\mathcal{M}_\gamma|}^2 &= \frac{1}{2}\sum_{espines}|\mathcal{M}_\gamma|^2, \\ &= \frac{e^4}{2q^4}\sum_{k_1,p_1}\sum_{k_2,p_2}[\bar{u}(k_1)\Gamma^\mu(l,q)u(p_1)\bar{u}(k_2)\gamma_\mu u(p_2)] \\ &\quad \times [\bar{u}(p_2)\gamma_\nu u(k_2)\bar{u}(p_1)\Gamma^\nu(l,-q)u(k_1)].\end{aligned} \quad (10)$$

After applying

$$\sum_{s=1}^{2} u_\alpha^{(s)}(p)\bar{u}_\beta^{(s)}(p) = (\not{p} + m_p)_{\alpha\beta}, \quad \text{particulas}, \quad (11)$$

$$\sum_{s=1}^{2} v_\alpha^{(s)}(p)\bar{v}_\beta^{(s)}(p) = (\not{p} - m_a)_{\alpha\beta}, \quad \text{antiparticulas}, \quad (12)$$

we obtain

$$\begin{aligned}\overline{|\mathcal{M}_\gamma|}^2 &= \frac{e^4}{q^4}Tr[\Gamma^\mu(l,q)(\not{p}_1 + m_\nu)\Gamma^\nu(l,-q)(\not{k}_1 + m_\nu)] \\ &\quad \times Tr[\gamma_\mu(\not{p}_2 + m_e)\gamma_\nu(\not{k}_2 + m_e)],\end{aligned}$$

$$= \frac{e^4}{q^4} N_\gamma^{\mu\nu} E_{\gamma\mu\nu}, \tag{13}$$

where $N^{\mu\nu}$ and $E_{\mu\nu}$ are given by

$$N_\gamma^{\mu\nu} = Tr[\Gamma^\mu(l,q)(\not{p}_1+m_\nu)\Gamma^\nu(l,-q)(\not{k}_1+m_\nu)], \tag{14}$$
$$E_{\gamma\mu\nu} = Tr[\gamma_\mu(\not{p}_2+m_e)\gamma_\nu(\not{k}_2+m_e)]. \tag{15}$$

Using the properties of the Dirac gamma matrices, we obtain

$$\begin{aligned}
E_{\gamma\mu\nu} &= 4[k_{2\nu}p_{2\mu}+k_{2\mu}p_{2\nu}+(m_e^2-k_2\cdot p_2)g_{\mu\nu}],\\
N_\gamma^{\mu\nu} &= \frac{1}{m_\nu^2}[F_Q^2(4m_\nu^2 k_1^\mu p_1^\nu+4m_\nu^2 k_1^\nu p_1^\mu-4m_\nu^2(k_1\cdot p_1)g^{\mu\nu}+4m_\nu^4 g^{\mu\nu})\\
&\quad+F_Q F_M(-2m_\nu^2 k_1^\nu q^\mu-2m_\nu^2 k_1^\mu q^\nu+2m_\nu^2 p_1^\nu q^\mu+2m_\nu^2 p_1^\mu q^\nu\\
&\quad+4m_\nu^2(k_1\cdot q)g^{\mu\nu}-4m_\nu^2(p_1\cdot q)g^{\mu\nu})\\
&\quad+F_M^2(-q^2 k_1^\nu p_1^\mu-q^2 k_1^\mu q^\nu+k_1^\nu(p_1\cdot q)q^\mu+k_1^\mu(p_1\cdot q)q^\nu+(k_1\cdot q)p_1^\nu q^\mu\\
&\quad+(k_1\cdot q)p_1^\mu q^\nu-(k_1\cdot p_1)q^\mu q^\nu+q^2(k_1\cdot p_1)g^{\mu\nu}-2(k_1\cdot q)(p_1\cdot q)g^{\mu\nu}-m_\nu^2 q^\mu q^\nu\\
&\quad+m_\nu^2 q^2 g^{\mu\nu})].
\end{aligned} \tag{16}$$

From Eq. (8), the square of the transition amplitude for the contribution of the Z boson of the SM is

$$\begin{aligned}
|\mathcal{M}_Z|^2 &= \mathcal{M}_Z \mathcal{M}_Z^\dagger,\\
&= 2G_F^2[\bar{u}(k_1)\gamma^\mu(g_{1V}^\nu-g_{1A}^\nu\gamma_5)u(p_1)]\\
&\quad\times[\bar{u}(k_2)\gamma_\mu(g_{1V}^e-g_{1A}^e\gamma_5)u(p_2)]\\
&\quad\times[\bar{u}(p_2)\gamma_\nu(g_{1V}^e-g_{1A}^e\gamma_5)u(k_2)]\\
&\quad\times[\bar{u}(p_1)\gamma^\nu(g_{1V}^\nu-g_{1A}^\nu\gamma_5)u(k_1)].
\end{aligned} \tag{17}$$

We add and average over the particle spin states since they are unobservable variables, obtaining

$$\begin{aligned}
\overline{|\mathcal{M}|^2} &= \frac{1}{2}\sum_{espines}|\mathcal{M}|^2\\
&= \frac{1}{2}\sum_{k_1,p_1}\sum_{k_2,p_2}2G_F^2[\bar{u}(k_1)\gamma^\mu(g_{1V}^\nu-g_{1A}^\nu\gamma_5)u(p_1)]\\
&\quad\times[\bar{u}(k_2)\gamma_\mu(g_{1V}^e-g_{1A}^e\gamma_5)u(p_2)][\bar{u}(p_2)\gamma_\nu(g_{1V}^e-g_{1A}^e\gamma_5)u(k_2)]\\
&\quad\times[\bar{u}(p_1)\gamma^\nu(g_{1V}^\nu-g_{1A}^\nu\gamma_5)u(k_1)].
\end{aligned} \tag{18}$$

Applying the matrix trace definition renders

$$\begin{aligned}\overline{|\mathcal{M}_Z|}^2 &= G_F^2 Tr[\gamma_\mu(g_{1V}^e - g_{1A}^e\gamma_5)(\not{p}_2 + m_e)\gamma_\nu(g_{1V}^e - g_{1A}^e\gamma_5)(\not{k}_2 + m_e)], \\ &\times Tr[\gamma^\mu(g_{1V}^\nu - g_{1A}^\nu\gamma_5)(\not{p}_1 + m_\nu)\gamma^\nu(g_{1V}^\nu - g_{1A}^\nu\gamma_5)(\not{k}_1 + m_\nu)], \\ &= G_F^2 N_Z^{\mu\nu} E_{Z\mu\nu}, \end{aligned} \quad (19)$$

with the following tensors

$$N_Z^{\mu\nu} = Tr[\gamma^\mu(g_{1V}^\nu - g_{1A}^\nu\gamma_5)(\not{p}_1 + m_\nu)\gamma^\nu(g_{1V}^\nu - g_{1A}^\nu\gamma_5)(\not{k}_1 + m_\nu)], \quad (20)$$
$$E_{Z\mu\nu} = Tr[\gamma_\mu(g_{1V}^e - g_{1A}^e\gamma_5)(\not{p}_2 + m_e)\gamma_\nu(g_{1V}^e - g_{1A}^e\gamma_5)(\not{k}_2 + m_e)]. \quad (21)$$

We develop the traces given in Eqs. (20) and (21) and substitute them in Eq. (19), obtaining

$$\begin{aligned}\overline{|\mathcal{M}_Z|}^2 &= 32 G_F^2\{(k_1 \cdot p_2)(k_2 \cdot p_1)([g_{1A}^e]^2([g_{1A}^\nu]^2 + [g_{1V}^\nu]^2) - 4[g_{1A}^e][g_{1A}^\nu][g_{1V}^e][g_{1V}^\nu] \\ &+ [g_{1V}^e]^2([g_{1A}^\nu]^2 + [g_{1V}^\nu]^2)) + m_e^2([g_{1A}^e]^2 - [g_{1V}^e]^2)([g_{1A}^\nu]^2 + [g_{1V}^\nu]^2)(k_1 \cdot p_1) \\ &+ [g_{1A}^e]^2[g_{1A}^\nu]^2(k_1 \cdot k_2)(p_1 \cdot p_2) + [g_{1A}^e]^2[g_{1A}^\nu]^2 m_\nu^2(k_2 \cdot p_2) + 2[g_{1A}^e]^2[g_{1A}^\nu]^2 m_e^2 m_\nu^2 \\ &+ [g_{1A}^e]^2[g_{1V}^\nu]^2(k_1 \cdot k_2)(p_1 \cdot p_2) - [g_{1A}^e]^2[g_{1V}^\nu]^2 m_\nu^2(k_2 \cdot p_2) \\ &- 2[g_{1A}^e]^2[g_{1V}^\nu]^2 m_e^2 m_\nu^2 + 4[g_{1A}^e][g_{1A}^\nu][g_{1V}^e][g_{1V}^\nu](k_1 \cdot k_2)(p_1 \cdot p_2) \\ &+ [g_{1A}^\nu]^2[g_{1V}^e]^2(k_1 \cdot k_2)(p_1 \cdot p_2) + [g_{1A}^\nu]^2[g_{1V}^e]^2 m_\nu^2(k_2 \cdot p_2) \\ &- 2[g_{1A}^\nu]^2[g_{1V}^e]^2 m_e^2 m_\nu^2 + [g_{1V}^e]^2[g_{1V}^\nu]^2(k_1 \cdot k_2)(p_1 \cdot p_2) \\ &- [g_{1V}^e]^2[g_{1V}^\nu]^2 m_\nu^2(k_2 \cdot p_2) + 2[g_{1V}^e]^2[g_{1V}^\nu]^2 m_e^2 m_\nu^2\}. \end{aligned} \quad (22)$$

In the 331RHν model, the couplings constants vector and axial-vector are equal and have the value $g_{1A}^\nu = g_{1V}^\nu$. Eq. (22) is hence reduced to

$$\begin{aligned}\overline{|\mathcal{M}_Z|}^2 &= 16 G_F^2 (2 g_{1A}^\nu)^2\{m_e^2[(g_{1A}^e)^2 - (g_{1V}^e)^2](k_1 \cdot p_1) \\ &+ [(g_{1A}^e) - (g_{1V}^e)]^2(k_1 \cdot p_2)(k_2 \cdot p_1) + [(g_{1A}^e) + (g_{1V}^e)]^2(k_1 \cdot k_2)(p_1 \cdot p_2)\}, \\ &= 16 G_F^2 (g_{1A}^\nu + g_{1V}^\nu)^2\{m_e^2[(g_{1A}^e)^2 - (g_{1V}^e)^2](k_1 \cdot p_1) \\ &+ [(g_{1A}^e) - (g_{1V}^e)]^2(k_1 \cdot p_2)(k_2 \cdot p_1) + [(g_{1A}^e) + (g_{1V}^e)]^2(k_1 \cdot k_2)(p_1 \cdot p_2)\}. \end{aligned} \quad (23)$$

The square of the transition amplitude for the contribution of the heavy gauge boson Z' of the 331RHν model is given by

$$
\begin{aligned}
\overline{|\mathcal{M}_{Z'}|}^2 &= 16 G_F^2 \left(\frac{M_Z}{M_{Z'}}\right)^2 (2g_{2A}^\nu)^2 \{m_e^2[(g_{2A}^e)^2 - (g_{2V}^e)^2](k_1 \cdot p_1) \\
&+ [(g_{2A}^e) - (g_{2V}^e)]^2 (k_1 \cdot p_2)(k_2 \cdot p_1) + [(g_{2A}^e) + (g_{2V}^e)]^2 (k_1 \cdot k_2)(p_1 \cdot p_2)\}, \\
&= 16 G_F^2 \left(\frac{M_Z}{M_{Z'}}\right)^2 (g_{2A}^\nu + g_{2V}^\nu)^2 \{m_e^2[(g_{2A}^e)^2 - (g_{2V}^e)^2](k_1 \cdot p_1) \quad (24) \\
&+ [(g_{2A}^e) - (g_{2V}^e)]^2 (k_1 \cdot p_2)(k_2 \cdot p_1) + [(g_{2A}^e) + (g_{2V}^e)]^2 (k_1 \cdot k_2)(p_1 \cdot p_2)\}.
\end{aligned}
$$

Finally, the terms of interference are

$$
\begin{aligned}
\overline{|\mathcal{M}_{int(\gamma-Z)}|}^2 &= \frac{4e^2}{q^2}\sqrt{2}[g_{1A}^\nu + g_{1V}^\nu] G_F \{-4F_Q g_{1A}^e (k_1 \cdot p_2)(k_2 \cdot p_1) \\
&+ 4F_Q g_{1A}^e (k_1 \cdot k_2)(p_1 \cdot p_2) + 4F_Q g_{1V}^e (k_1 \cdot p_2)(k_2 \cdot p_1) \\
&+ 4F_Q g_{1V}^e (k_1 \cdot k_2)(p_1 \cdot p_2) - 4F_Q g_{1V}^e m_e^2 (k_1 \cdot p_1) - 4F_Q g_{1V}^e m_\nu^2 (k_2 \cdot p_2) \\
&+ 8F_Q g_{1V}^e m_e^2 m_\nu^2 + 2F_M g_{1A}^e (k_1 \cdot p_2)(k_2 \cdot q) - 2F_M g_{1A}^e (k_1 \cdot k_2)(p_2 \cdot q) \\
&+ 2F_M g_{1A}^e (k_2 \cdot q)(p_1 \cdot p_2) - 2F_M g_{1A}^e (k_2 \cdot p_1)(p_2 \cdot q) \\
&+ F_M g_{1V}^e (k_1 \cdot q)(3m_e^2 - k_2 \cdot p_2) - F_M g_{1V}^e (k_1 \cdot p_2)(k_2 \cdot q) \quad (25) \\
&- F_M g_{1V}^e (k_1 \cdot k_2)(p_2 \cdot q) + F_M g_{1V}^e (k_2 \cdot q)(p_1 \cdot p_2) \\
&+ F_M g_{1V}^e (k_2 \cdot p_2)(p_1 \cdot q) + F_M g_{1V}^e (k_2 \cdot p_1)(p_2 \cdot q) \\
&- 3F_M g_{1V}^e m_e^2 (p_1 \cdot q)\},
\end{aligned}
$$

$$
\begin{aligned}
\overline{|\mathcal{M}_{int(\gamma-Z')}|}^2 &= \frac{4e^2}{q^2}\sqrt{2}[g_{2A}^\nu + g_{2V}^\nu] G_F \left(\frac{M_Z}{M_{Z'}}\right) \{-4F_Q g_{2A}^e (k_1 \cdot p_2)(k_2 \cdot p_1) \\
&+ 4F_Q g_{2A}^e (k_1 \cdot k_2)(p_1 \cdot p_2) + 4F_Q g_{2V}^e (k_1 \cdot p_2)(k_2 \cdot p_1) \\
&+ 4F_Q g_{2V}^e (k_1 \cdot k_2)(p_1 \cdot p_2) - 4F_Q g_{2V}^e m_e^2 (k_1 \cdot p_1) - 4F_Q g_{2V}^e m_\nu^2 (k_2 \cdot p_2) \\
&+ 8F_Q g_{2V}^e m_e^2 m_\nu^2 + 2F_M g_{2A}^e (k_1 \cdot p_2)(k_2 \cdot q) - 2F_M g_{2A}^e (k_1 \cdot k_2)(p_2 \cdot q) \\
&+ 2F_M g_{2A}^e (k_2 \cdot q)(p_1 \cdot p_2) - 2F_M g_{2A}^e (k_2 \cdot p_1)(p_2 \cdot q) \quad (26) \\
&+ F_M g_{2V}^e (k_1 \cdot q)(3m_e^2 - k_2 \cdot p_2) - F_M g_{2V}^e (k_1 \cdot p_2)(k_2 \cdot q) \\
&- F_M g_{2V}^e (k_1 \cdot k_2)(p_2 \cdot q) + F_M g_{2V}^e (k_2 \cdot q)(p_1 \cdot p_2) \\
&+ F_M g_{2V}^e (k_2 \cdot p_2)(p_1 \cdot q) + F_M g_{2V}^e (k_2 \cdot p_1)(p_2 \cdot q) \\
&- 3F_M g_{2V}^e m_e^2 (p_1 \cdot q)\},
\end{aligned}
$$

and

$$
\overline{|\mathcal{M}_{int(Z-Z')}|}^2 = 32 G_F^2 \left(\frac{M_Z}{M_{Z'}}\right)(2g_{1A}^\nu)(2g_{2A}^\nu)\{(g_{1A}^e - g_{1V}^e)(g_{2A}^e - g_{2V}^e)(k_1 \cdot p_2)(k_2 \cdot p_1)
$$

$$
\begin{aligned}
&+ (g^e_{1A} + g^e_{1V})(g^e_{2A} + g^e_{2V})(k_1 \cdot k_2)(p_1 \cdot p_2) \\
&+ m^2_e(k_1 \cdot p_1)(g^e_{1A}g^e_{2A} - g^e_{1V}g^e_{2V})\}, \\
&= 32 G^2_F \left(\frac{M_Z}{M_{Z'}}\right)(g^\nu_{1A} + g^\nu_{1V})(g^\nu_{2A} + g^\nu_{2V})\{(g^e_{1A} - g^e_{1V})(g^e_{2A} - g^e_{2V}) \\
&\times (k_1 \cdot p_2)(k_2 \cdot p_1) + (g^e_{1A} + g^e_{1V})(g^e_{2A} + g^e_{2V})(k_1 \cdot k_2)(p_1 \cdot p_2) \\
&+ m^2_e(k_1 \cdot p_1)(g^e_{1A}g^e_{2A} - g^e_{1V}g^e_{2V})\}.
\end{aligned}
\tag{27}
$$

2.3. Differential Cross-Section of $\nu_{\mu,\tau} - e^-$

The total differential cross-section of the reactions $\nu_{\mu,\tau} - e^-$ is given by

$$
\left(\frac{d\sigma}{d\Omega}\right)_{Total} = \left(\frac{d\sigma}{d\Omega}\right)_\gamma + \left(\frac{d\sigma}{d\Omega}\right)_Z + \left(\frac{d\sigma}{d\Omega}\right)_{Z'} + \left(\frac{d\sigma}{d\Omega}\right)_{int}.
\tag{28}
$$

This expression can then be written as a function of the energy of the electron recoil T

$$
\left(\frac{d\sigma}{dT}\right)_{total} = \left(\frac{d\sigma}{dT}\right)_\gamma + \left(\frac{d\sigma}{dT}\right)_Z + \left(\frac{d\sigma}{dT}\right)_{Z'} + \left(\frac{d\sigma}{dT}\right)_{int}.
\tag{29}
$$

where each contribution of Eq. (29) is given explicitly by

$$
\left(\frac{d\sigma(\nu - e)}{dT}\right)_\gamma = \frac{\frac{\pi \alpha^2 \mu^2_\nu}{m^2_e}\left[1 - \frac{T}{E_\nu}\right]}{T} + \frac{G^2_F m_e x^2}{2\pi}\left[1 - \frac{T}{E_\nu}\right]^2 + \frac{G^2_F m_e x^2}{2\pi}
$$
$$
- \frac{G^2_F x^2}{2\pi}\frac{m^2_e T}{E^2_\nu},
\tag{30}
$$

$$
\left(\frac{d\sigma(\nu - e)}{dT}\right)_Z = \frac{G^2_F m_e}{2\pi}[g^\nu_{1V} + g^\nu_{1A}]^2 \left[(g^e_{1V} + g^e_{1A})^2 + (g^e_{1V} - g^e_{1A})^2\left(1 - \frac{T}{E_\nu}\right)^2\right.
$$
$$
\left. + ((g^e_{1A})^2 - (g^e_{1V})^2)\frac{m_e T}{E^2_\nu}\right],
\tag{31}
$$

$$
\left(\frac{d\sigma(\nu - e)}{dT}\right)_{Z'} = \frac{G^2_F m_e}{2\pi}\left(\frac{M_Z}{M_{Z'}}\right)^2 [g^\nu_{2V} + g^\nu_{2A}]^2 \left[(g^e_{2V} + g^e_{2A})^2\right.
$$
$$
\left. + (g^e_{2V} - g^e_{2A})^2 \left(1 - \frac{T}{E_\nu}\right)^2 + ((g^e_{2A})^2 - (g^e_{2V})^2)\frac{m_e T}{E^2_\nu}\right],
$$

$$
\left(\frac{d\sigma(\nu - e)}{dT}\right)_{(\gamma - Z)} = \frac{G^2_F m_e x}{\pi}[g^\nu_{1V} + g^\nu_{1V}][(g^e_{1V} + g^e_{1A})
$$
$$
+ (g^e_{1V} - g^e_{1A})\left(1 - \frac{T}{E_\nu}\right)^2 - g^e_{1V}\frac{m_e T}{E^2_\nu}\right],
\tag{32}
$$

$$\left(\frac{d\sigma(\nu-e)}{dT}\right)_{(\gamma-Z')} = \frac{G_F^2 m_e x}{\pi}[g_{2V}^\nu + \tilde{g}_{2V}^\nu]\left(\frac{M_Z}{M_{Z'}}\right)[(g_{2V}^e + g_{2A}^e)$$
$$+ (g_{2V}^e - g_{2A}^e)\left(1-\frac{T}{E_\nu}\right)^2 - g_{2V}^e \frac{m_e T}{E_\nu^2}\right], \qquad (33)$$

$$\left(\frac{d\sigma(\nu-e)}{dT}\right)_{(Z-Z')} = \frac{G_F^2 m_e}{\pi}[g_{1V}^\nu + \tilde{g}_{1V}^\nu]$$
$$\times \ [g_{2V}^\nu + \tilde{g}_{2V}^\nu]\left(\frac{M_Z}{M_{Z'}}\right)[(g_{1V}^e + g_{1A}^e)(g_{2V}^e + g_{2A}^e)$$
$$+ (g_{1V}^e - g_{1A}^e)(g_{2V}^e - g_{2A}^e)\left(1-\frac{T}{E_\nu}\right)^2 \qquad (34)$$
$$- (g_{1A}^e g_{2A}^e - g_{1V}^e g_{2V}^e)\frac{m_e T}{E_\nu^2}\right].$$

In these expressions, G_F is the Fermi constant, μ_ν is the anomalous magnetic moment, $x = \frac{\sqrt{2}\pi\alpha\langle r^2\rangle}{3G_F}$ is the charge radius of the neutrino, m_e is the electron mass, T is its recoil energy, and E_ν is the neutrino energy, respectively. Furthermore, M_Z, $M_{Z'}$ are the gauge boson masses of the SM and the 331RHν model. It should be mentioned that the mixing angle ϕ, between $Z-Z'$, is contained in the coupling constants $g^e_{(1V,2V)}$, $g^\nu_{(1V,2V)}$, $g^e_{(1A,2A)}$ and $g^\nu_{(1A,2A)}$ given in Table 1.

Acknowledgments

We acknowledge support from CONACyT, SNI and PROFOCIE (México).

Conclusion

We have calculated the differential cross-section corresponding to the dispersion process $\nu_{\mu,\tau} - e^-$ in the context of the 331RHν model so that our results can be used in terrestrial and astrophysical experiments. The differential cross-section is in terms of the mass of the new gauge boson Z', the mixing angle ϕ, the magnetic moment of the neutrino μ_ν and the charge radius $\langle r^2 \rangle$. Furthermore, our results have been compared with those corresponding to the Standard Model in the decoupling limit $M_{Z'} \to \infty$ and $\phi = 0$. We have reproduced the results previously reported in the literature for the processes $\nu_{\mu,\tau} - e^-$ and in the context of the minimally extended Standard Model with neutrino electromagnetic properties [15].

References

[1] Glashow S. L., *Nucl. Phys.* **22**, 579 (1961).

[2] Weinberg S., *Phys. Rev. Lett.* **19**, 1264 (1967).

[3] Salam A., in *Elementary Particle Theory*, Ed. N. Svartholm (Almquist and Wiskell, Stockholm, 1968) 367.

[4] Giunti C. and Studenkin A., *Rev. Mod. Phys.* **87**, 531(2015).

[5] Fujikawa K. and Shrock R., *Phys. Rev. Lett.* **45**, 963 (1980).

[6] Shrock Robert E., *Nucl. Phys.* **B206**, 359 (1982).

[7] Christenson J. H., Cronin J. W., Fitch V. L., and Turlay R., *Phys. Rev. Lett.* **13**, 138 (1964).

[8] Abe K., et al., *Phys. Rev. Lett.* **87**, 091802 (2001).

[9] Aaij R., et al. [LHCb Collaboration], *J. High Energy Phys.* **07** (2014) 041.

[10] Raffelt G., *Stars as Laboratories for Fundamental Physics: The Astrophysics of Neutrinos, Axions, and Other Weakly Interacting Particles*, Chicago: University of Chicago Press, 1996.

[11] Pisano F. and Pleitez V., *Phys. Rev.* **D46**, 410 (1992).

[12] Frampton P. H., *Phys. Rev. Lett.* **69**, 2889 (1992).

[13] Foot R.,et al., *Phys. Rev.* **D50**, 34 (1994).

[14] Long H., *Phys. Rev.* **D53**, 437 (1996).

[15] Vogel P. and Engel J., *Phys. Rev.* **D39**, 3378 (1989).

[16] Bernabeu J., et al., *Phys. Rev.* **D62**, 113012 (2000).

[17] Bernabeu J., et al., *Phys. Rev. Lett.* **89**, 101802 (2000).

[18] Dvornikov M. S. and Studenikin A. I., *Jour. of Exp. and Theor. Phys.* **99**, 254 (2004).

[19] Broggini C., Giunti C., Studenikin A., *Adv. High Energy Phys.* **2012**, (2012) 459526.

In: Leptons
Editor: Christopher M. Villegas

ISBN: 978-1-53614-929-6
© 2019 Nova Science Publishers, Inc.

Chapter 4

TAU-LEPTON: DIPOLE MOMENTS IN THE B-L MODEL

A. Gutiérrez-Rodríguez[1,]* and M. A. Hernández-Ruíz[2]
[1]Unidad Académica de Física,
Universidad Autónoma de Zacatecas,
Zacatecas, México
[2]Unidad Académica de Ciencias Químicas,
Universidad Autónoma de Zacatecas,
Zacatecas, México

Abstract

We obtain analytical expressions for the total cross section of the process $e^+e^- \to \tau^+\tau^-\gamma$ in the context of the B-L model. The total cross section is in terms of the mass of the new gauge boson Z', the mixing angle θ' of the B-L model, the magnetic moment a_τ and the electric dipole moment d_τ of the τ-lepton.

1. Introduction

The Standard Model (SM) [1, 2, 3] of elementary particle physics is the theory that describes the interactions between the fundamental constituents of matter,

---

*E-mail address: alexgu@fisica.uaz.edu.mx.

that is, the strong, weak and electromagnetic interactions. Gravitational interaction is not contemplated because its effect is minimal on a microscopic scale. The SM is a quantum field theory gauge invariant under the symmetry group $SU(3)_C \times SU(2)_L \times U(1)_Y$. The theory is based on the identification of a set of fundamental constituents of matter that are sensitive to the aforementioned interactions. These constituents are the leptons and quarks (see Table 1). The fundamental constituents of the matter are structured in three generations or families and according to current experimental evidence, have no internal structure, form factors or excited states.

As mentioned above, the fundamental constituents of matter are structured in three generations or families. All the matter is constituted by particles of the first generation. The existence of the other two generations is one of the aspects that the SM does not fully explain and is one of the motivations for which studies are carried out on the properties of the particles of the second and third generation. This work focuses on the study of the electromagnetic properties of the charged lepton of the third generation, the τ-lepton.

Table 1. Fundamental constituents of the matter

Fermiones	I	II	III	Q
Quarks	u	c	t	2/3
	d	s	b	-1/3
Leptones	ν_e	ν_μ	ν_τ	0
	e	μ	τ	-1

The production processes of τ-lepton pairs $e^+e^- \to \tau^+\tau^-$, $e^+e^- \to \tau^+\tau^-\gamma$, $\gamma\gamma \to \tau^+\tau^-$, and $\gamma\gamma \to \tau^+\tau^-\gamma$, in high energy e^+e^- colliders have been used to set bounds on its electromagnetic and weak dipole moments [4, 5, 6, 7, 8, 9, 10, 11, 12, 13, 14]. In the SM [1, 2, 3], the τ anomalous magnetic moment (MM) $a_\tau = (g_\tau - 2)/2$ is predicted to be $(a_\tau)_{SM} = 0.0011773(3)$ [15, 16] and the respective electric dipole moment (EDM) d_τ is only generated by the GIM mechanism at very high order in the coupling constant [17]. Similarly, the weak MM and EDM are induced in the SM at the loop level giving $a_\tau^W = -(2.10 + 0.61i) \times 10^{-6}$ [18, 19] and $d_\tau^W \leq 8 \times 10^{-34}$ecm [20, 21]. Since the current bounds on these dipole moments [4, 5, 6, 7] are well above the SM predictions, it has been pointed out that these quantities are excellent candidates to look for physics beyond the SM [18, 19, 20, 21, 22, 23, 24, 25, 26, 27, 28, 29, 30]. The most general coupling of the photon to charged leptons, which is independent of the mode, may be parameterized in the following form:

$$\Gamma_\tau^\alpha = eF_1(q^2)\gamma^\alpha + \frac{ie}{2m_\tau}F_2(q^2)\sigma^{\alpha\mu}q_\mu$$
$$+ \frac{e}{2m_\tau}F_3(q^2)\sigma^{\alpha\mu}q_\mu\gamma_5$$
$$+ eF_4(q^2)\gamma_5(\gamma^\alpha - \frac{2q^\alpha m_\tau}{q^2}), \qquad (1)$$

where e is the charge of the electron, m_τ is the mass of the tau-lepton, $\sigma^{\alpha\mu} = \frac{i}{2}[\gamma^\alpha, \gamma^\mu]$ represents the spin 1/2 angular momentum tensor and $q = p' - p$ is the momentum transfer. In the static (classical) limit the q^2-dependent form factors $F_{1,2,3,4}(q^2)$ have familiar interpretations for $q^2 = 0$: $F_1(0) = Q_\tau$ is the electric charge; $F_2(0) = a_\tau$ its anomalous MM and $F_3(0) = \frac{2m_\tau}{e}d_\tau$ with d_τ its EDM. $F_4(q^2)$ is the Anapole form factor. The form factors corresponding to charge radius and the anapole moment are not considered in this work.

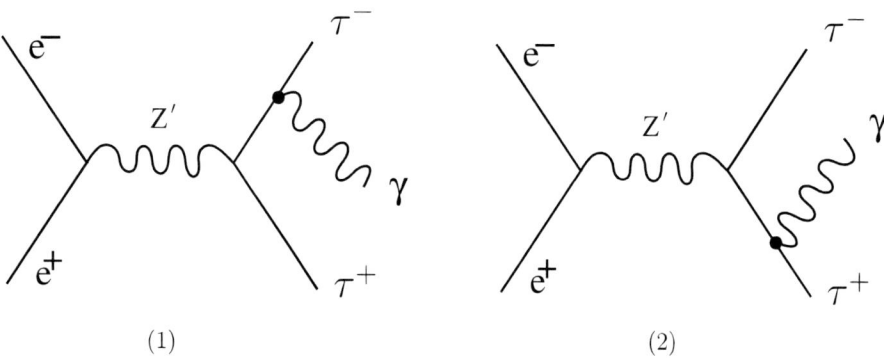

Figure 1. Feynman diagrams for the recation $e^+e^- \to \tau^+\tau^-\gamma$ in the B-L Model.

The latest bounds obtained for the electromagnetic dipole moments from the DELPHI and ALEPH collaborations at the 95% C.L. are: $-0.052 < a_\tau < 0.013$, $-0.22 < d_\tau(10^{-16} \text{ ecm}) < 0.45$ and $a_\tau^W < 1.1 \times 10^{-3}$, $d_\tau^W < 0.50 \times 10^{-17}$ecm [6, 7].

Our aim in this work is to analyze the reaction $e^+e^- \to \tau^+\tau^-\gamma$ in the Z' boson resonance where a large number of Z' events are collected at e^+e^- colliders in the context of the B-L model. We hope to constrain or eventually measure the electromagnetic and weak dipole moments of the τ by selecting $\tau^+\tau^-$ events accompanied by a hard photon. The Feynman diagrams which give the most important contribution to the cross section from $e^+e^- \to \tau^+\tau^-\gamma$ are shown in Fig. 1.

The $B-L$ model [31] is attractive due to its relatively simple theoretical structure, with the crucial test of the model being the detection of the new heavy neutral (Z') gauge boson.

This work is organized as follows: In Sect. I we present a very general description of the B-L model. In Sect. II we present the calculation of the cross section for the process $e^+e^- \to \tau^+\tau^-\gamma$. In Sect. III we present our conclusions.

2. The Lagrangian for the Gauge Sector

We consider an $SU(2)_L \times U(1)_Y \times U(1)_{B-L}$ model consisting of one doublet Φ and one singlet χ and briefly describe the lagrangian including the scalar, fermion and gauge sector. The Lagrangian for the gauge sector is given by [31, 32, 33, 34, 35]

$$\mathcal{L}_g = -\frac{1}{4}B_{\mu\nu}B^{\mu\nu} - \frac{1}{4}W_{\mu\nu}^a W^{a\mu\nu} - \frac{1}{4}Z'_{\mu\nu}Z'^{\mu\nu}, \qquad (2)$$

where $W_{\mu\nu}^a$, $B_{\mu\nu}$ and $Z'_{\mu\nu}$ are the field strength tensors for $SU(2)_L$, $U(1)_Y$ and $U(1)_{B-L}$, respectively.

In the Lagrangian of the $SU(2)_L \times U(1)_Y \times U(1)_{B-L}$ model, the terms for the interactions between neutral gauge bosons Z, Z' and a pair of fermions of the SM can be written in the form [32, 33, 31, 34]

$$\mathcal{L}_{NC} = \frac{-ig}{\cos\theta_W}\sum_f \bar{f}\gamma^\mu \frac{1}{2}(g_V^f - g_A^f \gamma^5) f Z_\mu$$
$$+ \frac{-ig}{\cos\theta_W}\sum_f \bar{f}\gamma^\mu \frac{1}{2}(g_V^{'f} - g_A^{'f} \gamma^5) f Z'_\mu. \quad (3)$$

From this Lagrangian we determine the expressions for the new couplings of the Z, Z' bosons with the SM fermions [34], which are given in Table 2.

Table 2. The new couplings of the Z, Z' bosons with the SM fermions in the B-L model. $g = e/\sin\theta_W$ and θ' is the $Z - Z'$ mixing angle

Particle	Couplings
$f\bar{f}Z$	$g_V^f = T_3^f \cos\theta' - 2Q_f \sin^2\theta_W \cos\theta' + \frac{2g_1'}{g}\cos\theta_W \sin\theta'$, $g_A^f = T_3^f \cos\theta'$
$f\bar{f}Z'$	$g_V^{'f} = -T_3^f \sin\theta' - 2Q_f \sin^2\theta_W \sin\theta' + \frac{2g_1'}{g}\cos\theta_W \cos\theta'$, $g_A^{'f} = -T_3^f \sin\theta'$

3. The Total Cross Section of the Reaction $e^+e^- \to \tau^+\tau^-\gamma$

In this section we calculate the total cross section for the reaction $e^+e^- \to \tau^+\tau^-\gamma$ using the neutral current lagrangian given in Eq. (3) for the B-L model for the Feynman diagrams 1 and 2 of Fig. 1. To determine the total cross-section in the resonance of the boson Z', we start from the Breit-Wigner resonance formula given by

$$\sigma(e^+e^- \to \tau^+\tau^-\gamma) = \frac{4\pi(2J+1)\Gamma_{e^+e^-}\Gamma_{\tau^+\tau^-\gamma}}{(s - M_{Z'}^2)^2 + M_{Z'}^2\Gamma_{Z'}^2}, \quad (4)$$

where $\Gamma_{e^+e^-}$ is the decay width of $Z' \to e^+e^-$ and $\Gamma_{\tau^+\tau^-\gamma}$ is the decay width of $Z' \to \tau^+\tau^-\gamma$, respectively.

3.1. The Decay Width of $Z' \to e^+e^-$

With the reaction

$$Z' \to e^+e^-, \qquad (5)$$

decay width is given for

$$\Gamma(Z' \to e^+e^-) = \frac{G_F M_{Z'}^3}{6\pi\sqrt{2}} \sqrt{1 - 4\frac{m_e^2}{M_{Z'}^2}} \left[(g_V'^e)^2 \left(1 + 2\frac{m_e^2}{M_{Z'}^2}\right) \right.$$

$$\left. + (g_A'^e)^2 \left(1 - 4\frac{m_e^2}{M_{Z'}^2}\right) \right], \qquad (6)$$

where G_F is the Fermi constant, m_e is the electron mass, $M_{Z'}$ is the heavy gauge boson mass of the $U(1)_{B-L}$ model and $g_V'^e (g_A'^e)$ are the couplings constant of $Z'e^+e^-$ which are given in Table 2.

3.2. The Decay Width of $Z' \to \tau^+\tau^-\gamma$

In this subsection we calculate the decay width of the reaction

$$Z' \to \tau^+\tau^-\gamma, \qquad (7)$$

using the neutral current Lagrangian given in Eq. (3) for the B-L model of the Feynman diagrams 1 and 2 of Fig. 1. The respective transition amplitudes are thus given by

$$\mathcal{M}_1 = \frac{-g}{2\cos\theta_W (l^2 - m_\tau^2)}$$
$$\times [\bar{u}(p_1)\Gamma_\tau^\alpha(\slashed{l} + m_\tau)\gamma^\beta(g_V'^\tau - g_A'^\tau\gamma_5)v(p_3)]\epsilon_\alpha^\lambda(\gamma)\epsilon_\beta^\lambda(Z'), \qquad (8)$$

$$\mathcal{M}_2 = \frac{-g}{2\cos\theta_W (k^2 - m_\tau^2)}$$
$$\times [\bar{u}(p_1)\gamma^\beta(g_V'^\tau - g_A'^\tau\gamma_5)(\slashed{k} + m_\tau)\Gamma_\tau^\alpha v(p_3)]\epsilon_\alpha^\lambda(\gamma)\epsilon_\beta^\lambda(Z'), \qquad (9)$$

where Γ_τ^α is the tau-lepton electromagnetic vertex which is defined in Eq. (1), while $\epsilon_\alpha^\lambda(\gamma)$ and $\epsilon_\beta^\lambda(Z')$ are the polarization vectors of the photon and the boson

Z', respectively. l (k) stands for the momentum of the virtual tau (antitau), and the coupling constants $g_V^{\prime f}$ and $g_A^{\prime f}$ with $f = e, \tau$ are given in Table 2.

After performing the corresponding Dirac algebra, we find that the MM, EDM and the parameters of the B-L model $g_V^{\prime \tau}$ and $g_A^{\prime \tau}$ give a contribution to the differential cross section for the process $e^+ e^- \to \tau^+ \tau^- \gamma$ in the form of:

$$\Gamma(Z' \to \tau^+ \tau^- \gamma) = \int \frac{\alpha}{12\pi^2 M_{Z'}^2 x_W (1 - x_W)} \left[\frac{e^2 F_2^2(q^2)}{4m_\tau^2} + \frac{e^2 F_3^2}{4m_\tau^2} \right]$$
$$\times \{[(g_V^{\prime \tau})^2 + (g_A^{\prime \tau})^2](s - 2\sqrt{s}E_\gamma) + (g_A^{\prime \tau})^2 E_\gamma^2 \sin^2 \theta_\gamma\}$$
$$\times E_\gamma dE_\gamma d\cos\theta_\gamma, \quad (10)$$

where $x_W \equiv \sin^2 \theta_W$, E_γ and $\cos\theta_\gamma$ are the energy and the opening angle of the emmited photon. Substituting the form factors $F_2(q^2)$ and $F_3(q^2)$ given in Eq. (1) we obtain

$$\Gamma(Z' \to \tau^+ \tau^- \gamma) = \int \frac{\alpha}{12\pi^2 M_{Z'}^2 x_W (1 - x_W)} \left[\frac{e^2 a_\tau^2}{4m_\tau^2} + d_\tau^2 \right]$$
$$\times \{[(g_V^{\prime \tau})^2 + (g_A^{\prime \tau})^2](s - 2\sqrt{s}E_\gamma) + (g_A^{\prime \tau})^2 E_\gamma^2 \sin^2 \theta_\gamma\}$$
$$\times E_\gamma dE_\gamma d\cos\theta_\gamma. \quad (11)$$

3.3. The Total Cross Section $e^+ e^- \to Z' \to \tau^+ \tau^- \gamma$

To determine the total cross section of the reaction $e^+ e^- \to \tau^+ \tau^- \gamma$, we substituted the decay widths of Eqs. (6) and (11) in the expression of the total cross section given for the Breit-Wigner resonance formula Eq. (4), resulting in

$$\sigma(e^+ e^- \to \tau^+ \tau^- \gamma) = \int \frac{\alpha^2 M_{Z'}}{12\pi M_Z^2} \left[\frac{e^2 a_\tau^2}{4m_\tau^2} + d_\tau^2 \right] \left[\frac{(g_V^{\prime e})^2 + (g_A^{\prime e})^2}{x_W^2 (1 - x_W)^2} \right]$$
$$\times \left[\frac{[(g_V^{\prime \tau})^2 + (g_A^{\prime \tau})^2](s - 2\sqrt{s}E_\gamma) + (g_A^{\prime \tau})^2 E_\gamma^2 \sin^2 \theta_\gamma}{[(s - M_{Z'}^2)^2 + M_{Z'}^2 \Gamma_{Z'}^2]} \right]$$
$$\times E_\gamma dE_\gamma d\cos\theta_\gamma, \quad (12)$$

where $M_{Z'}$ is the heavy gauge boson mass of the B-L model and $\Gamma_{Z'}$ is the total decay width of the Z' boson, respectively.

Acknowledgments

We acknowledge support from CONACyT, SNI and PROFOCIE (México).

Conclusion

We have calculated the total cross section corresponding to the reaction $e^+e^- \to Z' \to \tau^+\tau^-\gamma$ in the context of the $U(1)_{B-L}$ model. The cross section is in terms of the mass of the new gauge boson Z', the mixing angle θ', the magnetic moment of the tau a_τ and the electric dipole moment d_τ of the tau-lepton.

References

[1] Glashow, S. L., *Nucl. Phys.* **22**, 579 (1961).

[2] Weinberg, S., *Phys. Rev. Lett.* **19**, 1264 (1967).

[3] Salam, A., in *Elementary Particle Theory*, Ed. N. Svartholm (Almquist and Wiskell, Stockholm, 1968) 367.

[4] Lohmann, W., *Nucl. Phys. Proc. Suppl.* **144**, 122 (2005).

[5] L3 Collaboration, Achard, P., et al., *Phys. Lett.* **B585**, 53 (2004).

[6] DELPHI Collab., Abdallah, J., et al., *Eur. Phys. J.* **C35**, 159 (2004).

[7] ALEPH Collab., Heister, A., et al., *Eur. Phys. J.* **C30**, 291 (2003).

[8] Gutiérrez-Rodríguez, A., Hernández-Ruíz, M. A., and Luis-Noriega, L. N., *Mod. Phys. Lett.* **A19**, 2227 (2004).

[9] Gutiérrez-Rodríguez, A., Hernández-Ruíz, M. A., and Pérez, M. A., *Int. J. Mod. Phys.* **A22**, 3493 (2007).

[10] Gutiérrez-Rodríguez, A., *Mod. Phys. Lett.* **A25**, 703 (2010).

[11] Gutiérrez-Rodríguez, A., Hernández-Ruíz, M. A., Castañeda-Almanza, C. P., *J. Phys.* **G40**, 035001 (2013).

[12] Köksal, M., Inan, S. C., Billur, A. A., Bahar, M. K., Özgüven, Y., *arXiv*:1711.02405 [hep-ph].

[13] Özgüven, Y., Inan, S. C., Billur, A. A., Köksal, M., Bahar, M. K., *Nucl. Phys.* **B923**, 475 (2017).

[14] Billur A. A., Köksal M., *Phys. Rev. D89*, 037301 (2014).

[15] Samuel M. A., Li G. and Mendel R., *Phys. Rev. Lett.* **67**, 668 (1991); Erratum *ibid.* **69**, 995 (1992).

[16] Hamzeh F. and Nasrallah N. F., *Phys. Lett.* **B373**, 211 (1996).

[17] Barr S. M. and Marciano W. in *CP Violation*, ed. C. Jarlskog (World Scientific, Singapore, 1990).

[18] Bernabeu J. et al., *Nucl. Phys.* **B436**, 474 (1995).

[19] Bernabeu J. et al., *Phys. Lett.* **B326**, 168 (1994).

[20] Bernreuther W. et al. *Z. Phys.* **C43**, 117 (1989).

[21] Booth M. J., **hep-ph/9301293**.

[22] González-García M. C. and Novaes S. F., *Phys. Lett.* **B389**, 707 (1996).

[23] Poulose P. and Rindani S. D., **hep-ph/9708332**.

[24] Huang T., Lu W. and Tao Z., *Phys. Rev. D55*, 1643 (1997).

[25] Escribano R. and Massó E., *Phys. Lett.* **B395**, 369 (1997).

[26] Grifols J. A. and Méndez A., *Phys. Lett.* **B255**, 611 (1991); Erratum *ibid.* **B259**, 512 (1991).

[27] Taylor L., *Nucl. Phys. Proc. Suppl.* **B76**, 237 (1999).

[28] González-Sprinberg G. A., Santamaria A., Vidal J., *Int. Jour. Mod. Phys.* **A16** (Suppl.1B), 545 (2001).

[29] González-Sprinberg Gabriel A., Arcadi Santamaria, Jorge Vidal, *Nucl. Phys. Proc. Suppl.* **98**, 133 (2001).

[30] González-Sprinberg Gabriel A., A. Santamaria, J. Vidal, *Nucl. Phys.* **B582**, 3 (2000).

[31] Khalil S., *J. Phys. G: Nucl. Part. Phys.* **G35**, 055001 (2008).

[32] Ferroglia A., Lorca A. and van der Bij J. J., *Ann. Phys.* **16**, 563 (2007).

[33] Rizzo T. G., *arXiv*:hep-ph/0610104.

[34] Llamas-Bugarin A., Gutiérrez-Rodríguez A. and Hernández-Ruíz M. A. *Phys. Rev.* **D97**, 116008 (2017).

[35] Ramírez-Sánchez F., Gutiérrez-Rodríguez A. and Hernández-Ruíz M. A. *J. Phys.* **G43**, 095003 (2016).

BIBLIOGRAPHY

50 years of quarks

LCCN	2015288130
Type of material	Book
Main title	50 years of quarks / edited by Harald Fritzsch, Ludwig Maximilian University of Munich, Germany, Murray Gell-Mann, Santa Fe Institute, USA.
Published/Produced	Singapore; Hackensack, NJ: World Scientific, [2015]
	©2015
Description	x, 506 pages: illustrations (some color); 26 cm
ISBN	9789814618090 (hardcover)
	9814618098 (hardcover)
	9789814618106 (paperback)
	9814618101 (paperback)
LC classification	QC793.5.Q252 A15 2015
Variant title	Fifty years of quarks
Related names	Fritzsch, Harald, 1943- editor.
	Gell-Mann, Murray, editor.
Contents	A schematic model of baryons and mesons / M.

Gell-Mann -- Quarks / M. Gell-Mann -- Concrete quarks / G. Zweig -- On the way from sakatons to quarks / L.B. Okun -- My life with quarks / S.L. Glashow -- Quarks and the bootstrap era / D. Horn -- From symmetries to quarks and beyond / S. Meshkov -- How I got to work with Feynman on the covariant quark model / F. Ravndal -- What is a quark? / G.L. Kane & M.J. Perry -- Insights and puzzles in particle physics / H. Leutwyler -- Quarks and QCD / H. Fritzsch -- The discovery of gluon / J. Ellis -- Discovery of the gluon / S.L. Wu -- The Parton model and its applications / T.M. Yan & S.D. Drell -- From old symmetries to new symmetries: quark, leptons and B -- L / R.N. Mohapatra -- Quark mass hierarchy and flavor mixing puzzles / Z.-Z. Xing -- Analytical determination of the QCD quark masses / C. Dominquez -- CP violation in six quark scheme -- legacy of Sakata model / M. Kobayashi -- The constituent quark model -- nowadays / W. Plessas -- From [omega]⁻ to [omega][subscript b], doubly heavy baryons and exotics / M. Karliner -- Quark elastic scattering as a source of high transverse momentum mesons / R. Field -- Exclusive processes and the fundamental structure of hadrons / S.J. Brodsky -- Quark-gluon soup -- the perfectly liquid phase of QCD / U. Heinz -- Quarks and anomalies / R.J. Crewther -- Lessons from supersymmetry: "instead-of-confinement" mechanism / M. Shifman & A. Yung -- Quarks and a unified theory of nature fundamental forces / I. Antoniadis -- SU(8) family unification with boson-Fermion balance / S.L. Adler.

Subjects	Quarks.
	Quarks.
	Quark
Notes	Includes bibliographical references.

50 years of quarks

LCCN	2015506690
Type of material	Book
Main title	50 years of quarks / edited by Harald Fritzsch, Ludwig Maximilian University of Munich, Germany, Murray Gell-Mann, Santa Fe Institute, USA
Published/Produced	New Jersey: World Scientific, 2015.
	©2015
Description	x, 506 pages: illustrations (some color); 26 cm
ISBN	9789814618090
	9814618098
	9789814618106 (pbk)
	9814618101 (pbk)
LC classification	QC793.5.Q252 F54 2015
Variant title	Fifty years of quarks
Related names	Fritzsch, Harald, 1943- editor.
	Gell-Mann, Murray, editor.
Contents	A schematic model of baryons and mesons / M. Gell-Mann -- Quarks / M. Gell-Mann -- Concrete quarks / G. Zweig -- On the way from sakatons to quarks / L.B. Okun -- My life with quarks / S.L. Glashow -- Quarks and the bootstrap era / D. Horn -- From symmetries to quarks and beyond / S. Meshkov -- How I got to work with Feynman on the covariant quark model / F. Ravndal -- What is a quark? / G.L. Kane & M.J. Perry -- Insights and puzzles in particle physics / H. Leutwyler -- Quarks and QCD / H. Fritzsch --

	The discovery of gluon / J. Ellis -- Discovery of the gluon / S.L. Wu -- The Parton model and its applications / T.M. Yan & S.D. Drell -- From old symmetries to new symmetries: quark, leptons and B -- L / R.N. Mohapatra -- Quark mass hierarchy and flavor mixing puzzles / Z.-Z. Xing -- Analytical determination of the QCD quark masses / C. Dominquez -- CP violation in six quark scheme -- legacy of Sakata model / M. Kobayashi -- The constituent quark model -- nowadays / W. Plessas -- From [omega]' to [omega][subscript b], doubly heavy baryons and exotics / M. Karliner -- Quark elastic scattering as a source of high transverse momentum mesons / R. Field -- Exclusive processes and the fundamental structure of hadrons / S.J. Brodsky -- Quark-gluon soup -- the perfectly liquid phase of QCD / U. Heinz -- Quarks and anomalies / R.J. Crewther -- Lessons from supersymmetry: "instead-of-confinement" mechanism / M. Shifman & A. Yung -- Quarks and a unified theory of nature fundamental forces / I. Antoniadis -- SU(8) family unification with boson-Fermion balance / S.L. Adler.
Subjects	Quarks.
	Quarks.
	Quark
Notes	Includes bibliographical references.

Albert Einstein memorial lectures

LCCN	2011278911
Type of material	Book
Main title	Albert Einstein memorial lectures / editors, Jacob D. Bekenstein, Raphael Mechoulam.

Bibliography

Published/Created	Jerusalem: Israel Academy of Sciences and Humanities; Singapore; Hackensack, N.J.: World Scientific, c2012.
Description	ix, 204 p.: ill. (some col.); 24 cm.
ISBN	9789814329422 (hbk.)
	9814329428 (hbk.)
	9789814329439 (pbk.)
	9814329436 (pbk.)
LC classification	Q175 .A426 2012
Related names	Bekenstein, Jacob D., 1947-2015.
	Mechoulam, Raphael.
Summary	"This volume consists of a selection of the Albert Einstein Memorial Lectures presented annually at the Israel Academy of Sciences and Humanities. Delivered by eminent scientists and scholars, including Nobel laureates, they cover a broad spectrum of subjects in physics, chemistry, life science, mathematics, historiography and social issues. This distinguished memorial lecture series was inaugurated by the Israel Academy of Sciences and Humanities following an international symposium held in Jerusalem in March 1979 to commemorate the centenary of Albert Einstein's birth. Considering that Einstein's interests, activities and influence were not restricted to theoretical physics but spanned broad fields affecting society and the welfare of humankind, it was felt that these memorial lectures should be addressed to scientists, scholars and erudite laypersons rather than to physicists alone.."--publisher's description.
Contents	What can pure mathematics offer to society? / W. Timothy Gowers -- General covariance and the passive equations of physics / Shlomo

	Sternberg -- The structure of quarks and leptons / Haim Harari -- Beautiful theories / Steven Weinberg -- Harmless energy from nuclei / Carlo Rubbia -- Supramolecular chemistry: from molecular information toward self-organization and complex matter / Jean-Marie Lehn -- Chromatin and transcription / Roger Kornberg -- Energy, environment, and the responsibility of scientists / Yuan T Lee -- Res ipsa loquitur: history and mimesis / John E Wansbrough.
Subjects	Einstein, Albert, 1879-1955--Influence. Science--Philosophy. Science and the humanities

An introduction to particle physics and the standard model

LCCN	2009026775
Type of material	Book
Personal name	Mann, Robert, 1955-
Main title	An introduction to particle physics and the standard model / Robert Mann.
Published/Created	Boca Raton: CRC Press, c2010.
Description	xxi, 592 p.: ill.; 25 cm.
Links	Inhaltsverzeichnis. http://bvbm2.bib-bvb.de:8993 /F?func=service&doc_library=BVB01&doc_nu mber=018624995&line_number=0001&func_co de=DB_RECORDS&service_type=MEDIA
ISBN	9781420082982 (hardcover: alk. paper) 1420082981 (hardcover: alk. paper)
LC classification	QC793.2 .M36 2010
Summary	"An Introduction to the Standard Model of Particle Physics familiarizes readers with what is considered tested and accepted and in so doing, gives them a grounding in particle physics in general. Whenever possible, Dr. Mann takes an

historical approach showing how the model is linked to the physics that most of us have learned in less challenging areas. Dr. Mann reviews special relativity and classical mechanics, symmetries, conservation laws, and particle classification; then working from the tested paradigm of the model itself, he describes the standard model in terms of its electromagnetic, strong, and weak components; explores the experimental tools and methods of particle physics; introduces Feynman diagrams, wave equations, and gauge invariance, building up to the theory of quantum electrodynamics; describes the theories of the strong and electroweak interactions; and uncovers frontier areas and explores what might lie beyond our current concepts of the subatomic world."--Publisher's description.

Contents

Introduction and overview -- A review of special relativity -- Symmetries -- Conservation laws -- Particle classification -- Discrete symmetries -- Accelerators -- Detectors -- Scattering -- A toy theory -- Wave equations for elementary particles -- Gauge invariance -- Quantum electrodynamics -- Testing QED -- From nuclei to quarks -- The quark model -- Testing the quark model -- Heavy quarks and QCD -- From beta decay to weak interactions -- Charged leptonic weak interactions -- Charged weak interactions of quarks and leptons -- Electroweak unification -- Electroweak symmetry breaking -- Testing electroweak theory -- Beyond the standard model.

Subjects

Particles (Nuclear physics)

	Quark models.
	String models.
Notes	"A Taylor & Francis book."
	Includes bibliographical references and index.

Dynamics of the standard model

LCCN	2014006909
Type of material	Book
Personal name	Donoghue, John F., 1950- author.
Main title	Dynamics of the standard model / John F. Donoghue, University of Massachusetts, Eugene Golowich, University of Massachusetts, and Barry R. Holstein, University of Massachusetts.
Edition	Second edition.
Published/Produced	Cambridge, United Kingdom; New York: Cambridge University Press, [2014] ©2014
Description	xix, 573 pages: illustrations; 26 cm.
ISBN	9780521768672 (hardback) 0521768675 (hardback)
LC classification	QC794.6.S75 D66 2014
Related names	Golowich, Eugene, 1939- author. Holstein, Barry R., 1943- author.
Contents	Inputs to the Standard Model -- Interactions of the Standard Model -- Symmetries and anomalies -- Introduction to effective field theory -- Charged leptons -- Neutrinos -- Effective field theory for low-energy QCD -- Weak interactions of kaons -- Mass mixing and CP violation -- The Nc-1 expansion -- Phenomenological models -- Baryon properties -- Hadron spectroscopy -- Weak interactions of heavy quarks -- The Higgs boson -- The electroweak sector.

Bibliography

Subjects	Standard model (Nuclear physics)
	Electroweak interactions.
Notes	Includes bibliographical references (pages 545-566) and index.
Series	Cambridge monographs on particle physics, nuclear physics and cosmology; 35
	Cambridge monographs on particle physics, nuclear physics, and cosmology; 35.

Electroweak and strong interactions: phenomenology, concepts, models

LCCN	2011936749
Type of material	Book
Personal name	Scheck, Florian, 1936-
Main title	Electroweak and strong interactions: phenomenology, concepts, models / Florian Scheck.
Edition	3rd ed.
Published/Created	Heidelberg; New York: Springer, c2012.
Description	xx, 421 p.: ill.; 24 cm.
Links	Contributor biographical information http://www. loc.gov/catdir/enhancements/fy1312/2011936749-b.html
	Publisher description http://www.loc.gov/catdir/enhancements/fy1312/2011936749-d.html
	Table of contents only http://www.loc.gov/catdir/ enhancements/fy1312/2011936749-t.html
ISBN	9783642202407
LC classification	QC794 .S28 2012
Related names	Scheck, Florian, 1936- Leptons, hadrons, and nuclei.
Subjects	Nuclear reactions.
	Electroweak interactions.

	Particles (Nuclear physics)
	Strong interactions (Nuclear physics)
Notes	Rev. ed. of: Leptons, hadrons, and nuclei. 1983.
	Includes bibliographical references and index.
Series	Graduate texts in physics

Facts and mysteries in elementary particle physics

LCCN	2018006226
Type of material	Book
Personal name	Veltman, Martinus, author.
Main title	Facts and mysteries in elementary particle physics / Martinus Veltman, MacArthur Emeritus Professor of Physics, University of Michigan, Ann Arbor, USA and NiKHEF, Amsterdam, the Netherlands.
Edition	Revised edition.
Published/Produced	New Jersey: World Scientific, [2018]
Description	viii, 344 pages: illustrations (some color); 24 cm
ISBN	9789813237056 (hard cover; alk. paper)
	9813237058 (hard cover; alk. paper)
	9789813237490 (pbk; alk. paper)
	981323749X (pbk; alk. paper)
LC classification	QC793.2 .V45 2018
Summary	"This book provides a comprehensive overview of modern particle physics accessible to anyone with a true passion for wanting to know how the universe works. We are introduced to the known particles of the world we live in. An elegant explanation of quantum mechanics and relativity paves the way for an understanding of the laws that govern particle physics. These laws are put into action in the world of accelerators, colliders and detectors found at institutions such as CERN and Fermilab that are in the forefront of technical

innovation. Real world and theory meet using Feynman diagrams to solve the problems of infinities and deduce the need for the Higgs boson. Facts and Mysteries in Elementary Particle Physics offers an incredible insight from an eyewitness and participant in some of the greatest discoveries in 20th century science. From Einstein's theory of relativity to the spectacular discovery of the Higgs particle, this book will fascinate and educate anyone interested in the world of quarks, leptons and gauge theories. This book also contains many thumbnail sketches of particle physics personalities, including contemporaries as seen through the eyes of the author. Illustrated with pictures, these candid sketches present rare, perceptive views of the characters that populate the field. The Chapter on Particle Theory, in a pre-publication, was termed "superbly lucid" by David Miller in Nature (Vol. 396, 17 Dec. 1998, p. 642)"-- Provided by publisher.

Subjects	Particles (Nuclear physics)
Notes	Includes bibliographical references and indexes.

From photons to Higgs: a story of light

LCCN	2013045018
Type of material	Book
Personal name	Han, M. Y., author.
Uniform title	Story of light
Main title	From photons to Higgs: a story of light / Moo-Young Han, Korea Advanced Institute of Science and Technology (KAIST), Korea Et Duke University, USA.
Edition	2nd edition.

Published/Produced	New Jersey: World Scientific, [2014] ©2014
Description	xiv, 127 pages: illustrations; 24 cm
ISBN	9789814579957 (hardcover: alk. paper)
	9814579955 (hardcover: alk. paper)
	9789814583862 (softcover: alk. paper)
	9814583863 (softcover: alk. paper)
LC classification	QC793.2 .H36 2014
Variant title	Story of light
Contents	Particles and fields I: Dichotomy -- Lagangian and Hamiltonian dynamics -- Canonical quantization -- Particles and fields II: Duality -- Equations for duality -- Electromagnetic field -- Emulation of light I: Matter fields -- Road map for field quantization -- Particles and fields III: Particles as quanta of fields -- Emulation of light II: Interactions -- Triumph and wane -- Leptons and quarks -- What is gauge field theory? -- The weak gauge fields -- The Higgs mechanism and the electroweak gauge fields -- The Higgs particle -- Evolution of the strong force -- The history of color SU(3) symmetry -- Quantum chromodynamics, QCD -- Appendix 1: The natural unit system -- Appendix 2: Notation -- Appendix 3: Velocity-dependent potential -- Appendix 4: Fourier decomposition of field -- Appendix 5: Mass units for particles -- Appendix 6. Mass-range relation.
Subjects	Particles (Nuclear physics)
	Quantum field theory.
Notes	Revision of: A story of light. 2004.
	Includes bibliographical references and index.

Gauge theories of the strong, weak, and electromagnetic interactions

LCCN	2013020713
Type of material	Book
Personal name	Quigg, Chris, author.
Main title	Gauge theories of the strong, weak, and electromagnetic interactions / Chris Quigg.
Edition	Second edition.
Published/Produced	Princeton, New Jersey: Princeton University Press, [2013]
Description	xiii, 482 pages; 26 cm
ISBN	9780691135489 (hardback)
LC classification	QC793.3.G38 Q5 2013
Summary	"This completely revised and updated graduate-level textbook is an ideal introduction to gauge theories and their applications to high-energy particle physics, and takes an in-depth look at two new laws of nature--quantum chromodynamics and the electroweak theory. From quantum electrodynamics through unified theories of the interactions among leptons and quarks, Chris Quigg examines the logic and structure behind gauge theories and the experimental underpinnings of today's theories. Quigg emphasizes how we know what we know, and in the era of the Large Hadron Collider, his insightful survey of the standard model and the next great questions for particle physics makes for compelling reading. The brand-new edition shows how the electroweak theory developed in conversation with experiment. Featuring a wide-ranging treatment of electroweak symmetry breaking, the physics of the Higgs boson, and the importance of the 1-TeV scale, the book moves

	beyond established knowledge and investigates the path toward unified theories of strong, weak, and electromagnetic interactions. Explicit calculations and diverse exercises allow readers to derive the consequences of these theories. Extensive annotated bibliographies accompany each chapter, amplify points of conceptual or technical interest, introduce further applications, and lead readers to the research literature. Students and seasoned practitioners will profit from the text's current insights, and specialists wishing to understand gauge theories will find the book an ideal reference for self-study. Brand-new edition of a landmark text introducing gauge theories Consistent attention to how we know what we know Explicit calculations develop concepts and engage with experiment Interesting and diverse problems sharpen skills and ideas Extensive annotated bibliographies "-- Provided by publisher.
Subjects	Gauge fields (Physics)
	Nuclear reactions.
	Strong interactions (Nuclear physics)
	Science / Quantum Theory.
	Science / Electromagnetism.
Notes	Includes bibliographical references and indexes.

How did we all begin: where is God in all that?

LCCN	2010016708
Type of material	Book
Personal name	Kalman, C. S. (Calvin S.)
Main title	How did we all begin: where is God in all that? / author, Calvin S. Kalman.
Published/Created	Hauppauge, N.Y.: Nova Science Publishers,

	2010.
Description	xvi, 138 p.: ill. (some col.); 23 cm.
ISBN	9781616683641 (softcover)
LC classification	QB981 .K135 2010
Contents	Our view of the universe changes -- The end of classical determinism -- Remarkable discoveries at at the end of the 19th century -- The quantum hypothesis -- Reading the book of nature -- The modern atom -- The particle zoo -- Neutrinos and supernovae -- Quarks and leptons -- Symmetries and fields -- How does the force between particles take place through the exchange of photons or gluons? -- On galaxies -- The universe is expanding -- Evidence for the big bang -- Before the big bang -- Will the universe end in a big crunch -- Dark matter.
Subjects	Cosmology.
	God--Proof, Cosmological.
Notes	Includes index.

Introduction to elementary particle physics

LCCN	2013046728
Type of material	Book
Personal name	Bettini, Alessandro, 1939- author.
Main title	Introduction to elementary particle physics / Alessandro Bettini, University of Padua, Italy.
Edition	Second edition.
Published/Produced	New York: Cambridge University Press, [2014] ©2014
Description	xvii, 474 pages: illustrations; 26 cm
ISBN	9781107050402 (hardback)
LC classification	QC794.6.S75 B48 2014
Contents	Preliminary notions -- Nucleons, leptons and mesons -- Symmetries -- Hadrons -- Quantum

	electrodynamics -- Chromodynamics -- Weak interactions -- The neutral mesons oscillations and CP violation -- The standard model -- Neutrinos -- Epilogue.
Subjects	Standard model (Nuclear physics)
	Standard model (Nuclear physics)--Problems, exercises, etc.
Notes	Includes bibliographical references (pages 458-465) and index.

Leptons and quarks: special edition commemorating the discovery of the Higgs boson

LCCN	2014451004
Type of material	Book
Personal name	Okun', L. B. (Lev Borisovich), author.
Uniform title	Leptony i kvarki. English
Main title	Leptons and quarks: special edition commemorating the discovery of the Higgs boson / Lev B. Okun, Institute of Theoretical and Experimental Physics (ITEP), Russia; translated by V.I. Kisin.
Edition	Special edition.
Published/Produced	Singapore; Hackensack, NJ: World Scientific, [2014]
	©2014
Description	xix, 409 pages: illustrations; 26 cm
ISBN	9789814603003
	9814603007
LC classification	QC793.5.L422 O3813 2014
Related names	Kisin, V. I., translator.
Summary	"The book "Leptons and Quarks" was first published in the early 1980s, when the program of the experimental search for the intermediate bosons W and Z and Higgs boson H was

Bibliography

formulated. The aim and scope of the present extended edition of the book, written after the experimental discovery of the Higgs boson in 2012, is to reflect the various stages of this 30+ years search. Along with the text of the first edition of "Leptons and Quarks" it contains extracts from a number of books published by World Scientific and an article from "On the concepts of vacuum and mass and the search for higgs" available from http://www.worldscientific. com/worldscinet/ mpla or from http://arxiv.org/ abs/1212.1031. The book is unique in communicating the Electroweak Theory at a basic level and in connecting the concept of Lorenz invariant mass with the concept of the Extended Standard Model, which includes gravitons as the carriers of gravitational interaction."--Page 4 of cover.

Contents Introduction -- Structure of weak currents -- Muon decay -- Strangeness-conserving leptonic decays of hadrons. Properties of the ud-current -- Leptonic decays of pions and nucleons -- Leptonic decays of K-mesons and hyperons -- Strangeness-changing non-leptonic interactions -- Phenomenology of non-leptonic decays of hyperons -- Dynamics of non-leptonic decays of hyperons -- Non-leptonic decays of K-mesons -- Neutral K-mesons in vacuum and in matter -- Violation of CP invariance -- Decays of the [symbol]-lepton -- Decays of charmed particles -- Weak decays of b- and t-quarks -- Neutrino-electron interactions -- Neutrino-nucleon interactions -- Renormalizability -- Gauge invariance -- Spontaneous symmetry breaking --

	Standard model of the electroweak interaction -- Neutral currents -- Properties of intermediate bosons -- Properties of Higgs bosons -- Grand unification -- Superunification -- Particles and the universe.
Subjects	Leptons (Nuclear physics)
	Quarks.
	Weak interactions (Nuclear physics)
Notes	Includes bibliographical references and index.

Massive neutrinos: flavor mixing of leptons and neutrino oscillations

LCCN	2015026505
Type of material	Book
Main title	Massive neutrinos: flavor mixing of leptons and neutrino oscillations / editor, Harald Fritzsch, Ludwig Maximilian University of Munich, Germany.
Published/Produced	Singapore; Hackensack, NJ: World Scientific, [2015]
	©2015
Description	xi, 294 pages: illustrations (some color); 26 cm
ISBN	9789814704762 (hardcover: alk. paper)
	9814704768 (hardcover: alk. paper)
LC classification	QC793.5.N42 M37 2015
Related names	Fritzsch, Harald, 1943- editor.
Contents	Birth of lepton flavor mixing / M. Kobayashi -- Neutrino masses and flavor mixing / H. Fritzsch -- Fermion mass matrices, textures and beyond / M. Gupta, P. Fakay, S. Sharma & G. Ahuja -- Status and implications of neutrino masses: a brief panorama / J.W.F. Valle -- Neutrino masses and So10 unification / P. Minkowski -- Relating small neutrino masses and mixing / A.

Raychaudhuri -- Predictions for the Dirac CP violation phase in the neutrino mixing matrix / S.T. Petcov, J. Girardi & A.V. Titov -- Sterile neutrinos in E_6 / J.L. Rosner -- Phenomenology of light sterile neutrinos / C. Giunti -- Neutrino-antineutrino mass splitting in SM / K. Fujikawa -- The strong CP problem and discrete symmetries / M. Spinrath -- Neutrino interaction with background matter in a noninertial frame / M. Dvornikov -- Seesaw models with minimal flavor violation / X.G. He -- Generating Majorana neutrino masses with loops / C.Q. Geng -- Three-neutrino oscillation parameters: status and prospects / F. Capozzi, G.L. Fogli, E. Lisi, A. Marrone, D. Montanino & A. Palazzo -- On the Majorana neutrinos and neutrinoless double beta decays / Z.Z. Xing & Y.L. Zhou -- Dirac or inverse seesaw neutrino masses from gauged B-L symmetry / E. Ma & R. Srivastava -- Searching for radiative neutrino mass generation at the LHC / R.R. Volkas -- Lepton-flavor violating signatures in supersymmetric U(1)' seesaw / E.J. Chun -- From electromagnetic neutrinos to new electromagnetic radiation mechanism in neutrino fluxes / I Balantsev & A. Studenikin -- Lepton number violation and the baryon asymmetry of the universe / J. Harz, W.C. Huang & H. Paes -- Status of the Majorana demonstrator: a search for neutrinoless double-beta decay / Y. Efremenko -- Towards neutrino mass spectroscopy using atoms/molecules / M. Yoshimura -- Detection prospects of the cosmic neutrino background / Y.F. Li -- Supernova bounds on keV-mass sterile neutrinos / S. Zhou -

98   *Bibliography*

	- Precision calculations for supersymmetric Higgs bosons / W. Hollik -- Minimal supersymmetric standard model with gauged baryon and lepton numbers / B. Fornal -- From the fourth color to spin-charge separation -- neutrinos and spinons / X. Chi.
Subjects	Neutrinos.
	Lepton interactions.
	Quantum flavor dynamics.
	Leptons (Nuclear physics)
Notes	Includes bibliographical references and index.
Series	Advanced series on directions in high energy physics; vol. 25
	Advanced series on directions in high energy physics; v. 25.

Modern elementary particle physics

LCCN	2016048277
Type of material	Book
Personal name	Kane, G. L., author.
Main title	Modern elementary particle physics / Gordon Kane, University of Michigan, Ann Arbor.
Edition	Second edition.
Published/Produced	Cambridge, United Kingdom; New York, NY, USA: Cambridge University Press, 2017.
	©2017
Description	xiv, 226 pages; 25 cm
ISBN	9781107165083 (hardback; alk. paper)
	1107165083 (hardback; alk. paper)
LC classification	QC793.2 .K36 2017
Summary	"Written for students and scientists wanting to learn about the Standard Model of particle physics. Only an introductory course knowledge about quantum theory is needed. This text

provides a pedagogical description of the theory, and incorporates the recent Higgs boson and top quark discoveries. With its clear and engaging style, this new edition retains its essential simplicity. Long and detailed calculations are replaced by simple approximate ones. It includes introductions to accelerators, colliders, and detectors. Several main experimental tests of the Standard Model are explained. Descriptions of some well-motivated extensions of the Standard Model prepares the reader for new developments. It emphasizes the concepts of gauge theories and Higgs physics, electroweak unification and symmetry breaking, and how force strengths vary with energy, provides a solid foundation for those working in the field, and for those who simply want to learn about the Standard Model"-- Provided by publisher.

Contents Relativistic notation, Lagrangians, and interactions -- Gauge invariance -- Non-abelian gauge theories -- Dirac notation for spin -- The Standard Model Lagrangian -- The electroweak theory and quantum chromodynamics -- Masses and the Higgs mechanism -- Cross sections, decay widths, and lifetimes; W and Z decays -- Production and properties of W± and Z^0-- Measurement of electroweak and QCD parameters; the muon lifetime -- Accelerators - present and future -- Experiments and detectors -- Low energy and non-accelerator experiments -- Observation of the Higgs boson at the CERN LHC; is it the Higgs boson? -- Colliders and tests of the Standard Model; particles are point-like -- Quarks and gluons, confinement and jets --

Bibliography

	Hadrons, heavy quarks, and strong isospin invariance -- Coupling strengths depend on momentum transfer and on virtual particles -- Quark (and lepton) mixing angles -- CP violation -- Overview of physics beyond the Standard Model -- Grand unification -- Neutrino masses -- Dark matter -- Supersymmetry.
Subjects	Particles (Nuclear physics)
	Standard model (Nuclear physics)
	Quarks.
	Leptons (Nuclear physics)
Notes	Includes bibliographical references and index.

Modern particle physics

LCCN	2013002757
Type of material	Book
Personal name	Thomson, Mark, 1966-
Main title	Modern particle physics / Mark Thomson, University of Cambridge.
Published/Produced	Cambridge, United Kingdom; New York: Cambridge University Press, 2013.
Description	xvi, 554 pages: illustrations; 26 cm
Links	Cover image http://assets.cambridge.org/97811070/34266/cover/9781107034266.jpg
ISBN	9781107034266 (hardback)
	1107034264 (hardback)
LC classification	QC793.2 .T46 2013
Summary	"Unique in its coverage of all aspects of modern particle physics, this textbook provides a clear connection between the theory and recent experimental results, including the discovery of the Higgs boson at CERN. It provides a comprehensive and self-contained description of the Standard Model of particle physics suitable

Bibliography

for upper-level undergraduate students and graduate students studying experimental particle physics. Physical theory is introduced in a straightforward manner with full mathematical derivations throughout. Fully-worked examples enable students to link the mathematical theory to results from modern particle physics experiments. End-of-chapter exercises, graded by difficulty, provide students with a deeper understanding of the subject. Online resources available at www.cambridge.org/MPP feature password-protected fully-worked solutions to problems for instructors, numerical solutions and hints to the problems for students and PowerPoint slides and JPEGs of figures from the book"-- Provided by publisher.

Contents   Machine generated contents note: 1. Introduction; 2. Underlying concepts; 3. Decay rates and cross sections; 4. The Dirac equation; 5. Interaction by particle exchange; 6. Electron-positron annihilation; 7. Electron-proton elastic scattering; 8. Deep inelastic scattering; 9. Symmetries and the quark model; 10. Quantum chromodynamics; 11. The weak interaction; 12. The weak interactions of leptons; 13. Neutrinos and neutrino oscillations; 14. CP violation and weak hadronic interactions; 15. Electroweak unification; 16. Tests of the Standard Model; 17. The Higgs boson; 18. The Standard Model and beyond; Appendixes; References; Further reading; Index.

Subjects   Particles (Nuclear physics)--Textbooks.
Science / Nuclear Physics.

Notes	Includes bibliographical references (page 545) and index.

New physics at the Large Hadron Collider

LCCN	2016044199
Type of material	Book
Meeting name	Conference on New Physics at the Large Hadron Collider (2016: Nanyang Technological University), author.
Main title	New physics at the Large Hadron Collider / editor Harald Fritzsch, University of Munich, Germany.
Published/Produced	New Jersey: World Scientific, [2017]
Description	xi, 444 pages: illustrations; 25 cm
ISBN	9789813145498 (hardcover)
	9813145498 (hardcover)
LC classification	QC770 .C58 2017
Related names	Fritzsch, Harald, 1943- editor.
Summary	"The Standard Theory of Particle Physics describes successfully the observed strong and electroweak interactions, but it is not a final theory of physics, since many aspects are not understood: (1) How can gravity be introduced in the Standard Theory? (2) How can we understand the observed masses of the leptons and quarks as well as the flavor mixing angles? (3) Why are the masses of the neutrinos much smaller than the masses of the charged leptons? (4) Is the new boson, discovered at CERN, the Higgs boson of the Standard Theory or an excited weak boson? (5) Are there new symmetries at very high energy, e.g., a broken supersymmetry? (6) Are the leptons and quarks point-like or composite particles? (7) Are the

leptons and quarks at very small distances one-dimensional objects, e.g., superstrings? This proceedings volume comprises papers written by the invited speakers discussing the many important issues of the new physics to be discovered at the Large Hadron Collider"--Provided by publisher.

Contents Accelerator considerations of large circular colliders / Alex Chao -- Physics potential and motivations for a muon collider / Mario Greco -- Pentaquarks and possible anomalies at LHCb / George Laferty -- Neutrino masses and SO10 unification / P. Minkowski -- Neutrino experiments: hierarchy, CP, CPT / Manmohan Gupta, Monika Randhawa and Mandip Singh -- Constraining the texture mass matrices / Gulsheen Ahuja -- Rare B-meson decays at the crossroads / Ahmed Ali -- Exploring the standard model at the LHC / Brigitte Vachon -- Meson/baryon/tetraquark supersymmetry from superconformal algebra and light-front holography / Stanley J. Brodsky, Guy F. de Teramond, Hans Gunter Dosch and Cedric Lorce -- The spin-charge-family theory / Norma Susana Mankoc Borstnik -- Search for direct CP violation in baryonic b-hadron decays / C.Q. Geng and Y.K. Hsiao -- New physics and astrophysical neutrinos in IceCube / Atsushi Watanabe -- The 750 GeV diphoton excess and SUSY / S. Heinemeyer -- Constraints on the [omega pi] form factor from analyticity and unitarity / B. Ananthanarayan, Irinel Caprini and Bastian Kubis -- Dynamical tuning of the initial condition in small field inflations-can we testify the CW mechanism in the universe / Satoshi Iso -- Physics of Higgs boson family / Ngee-Pong Chang -- On the breaking of mu-tau flavor symmetry / Zhen-Hua Zhao -- Neutrino mass ordering in future neutrinoless double beta decay experiments / Jue Zhang -- Predicting the CP-phase for neutrinos / Eiichi Takasugi -- Sum rules for leptons / Martin Spinrath -- Composite weak bosons at the Large Hadron Collider / Harald Fritzsch -- Searching for composite Higgs models at the LHC / Thomas Flacke -- Gauge-Higgs EW and grand unification / Yutaka Hosotani -- Colour octet extension of 2HDM / German Valencia -- New physics/resonances in vector boson scattering at the LHC / Jurgen Reuter, Wolfgang Kilian, Thorsten Ohl and Marco Sekulla -- Dimensional regularization is generic / Kazuo Fujikawa -- A de-gauging approach to physics beyond the standard model / Chi Xiong -- Extension of standard model in multi-spinor field formalism - visible and dark sectors / Ikuo S. Sogami -- Aspects of string phenomenology and new physics / I. Antoniadis -- Cosmological constant vis-a-vis dynamical vacuum: bold challenging the LCDM / Joan Sola.

Subjects Nuclear physics--Congresses.
Particles (Nuclear physics)--Congresses.

Notes Proceedings of the Conference on New Physics at the Large Hadron Collider, held at the Nanyang Technological University, Singapore, February 26-March 4, 2016.

Quarks, leptons and the big bang

LCCN	2016023876
Type of material	Book
Personal name	Allday, Jonathan, author.
Main title	Quarks, leptons and the big bang / Jonathan Allday.
Edition	Third edition.
Published/Produced	Boca Raton, FL: CRC Press, Taylor & Francis Group, [2017]
Description	xviii, 375 pages: illustrations; 25 cm
ISBN	9781498773119 (paperback; alk. paper)
	1498773117 (pbk.; alk. paper)
	(e-book)
	(e-book)
LC classification	QC793.2 .A394 2017
Related names	Allday, Jonathan, author.
Summary	"Quarks, Leptons and The Big Bang, Third Edition, is a clear, readable and self-contained introduction to particle physics and related areas of cosmology. It bridges the gap between non-technical popular accounts and textbooks for advanced students. The book concentrates on presenting the subject from the modern perspective of quarks, leptons and the forces between them. This book will be of interest to students, teachers and general science readers interested in fundamental ideas of modern physics. This edition brings the book completely up to date by including advances in particle physics and cosmology, such as the discovery of the Higgs boson, the LIGO gravitational wave discovery and the WMAP and PLANCK results. FEATURES Builds on the success of previous editions, which have received very

	complimentary reviews Focuses on key ideas as they are now, rather than taking a historical approach Restricted use of mathematics, making the book suitable for high school or first-year university level This third edition incorporates advances in particle physics and cosmology Written in an accessible yet rigorous style"-- Provided by publisher.
Contents	The standard model -- Relativity for particle physics -- Quantum theory -- The leptons -- Antimatter -- Hadrons -- Hadron reactions -- Particle decays -- The evidence for quarks -- Experimental techniques -- Exchange forces -- The big bang -- The geometry of space.
Subjects	Particles (Nuclear physics)
	Cosmology.
	Big bang theory.
Notes	Includes bibliographical references index.

Revolutions in twentieth-century physics

LCCN	2012030789
Type of material	Book
Personal name	Griffiths, David J. (David Jeffery), 1942-
Main title	Revolutions in twentieth-century physics / David J. Griffiths, Reed College.
Published/Produced	Cambridge, UK: Cambridge University Press, 2013.
Description	x, 174 pages: ill.; 23 cm
Links	Cover image http://assets.cambridge.org/97811076/02175/cover/9781107602175.jpg
ISBN	9781107602175 (pbk.)
LC classification	QC174.12 .G75154 2013
Summary	"The conceptual changes brought by modern physics are important, radical and fascinating,

yet they are only vaguely understood by people working outside the field. Exploring the four pillars of modern physics - relativity, quantum mechanics, elementary particles and cosmology - this clear and lively account will interest anyone who has wondered what Einstein, Bohr, Schrödinger and Heisenberg were really talking about. The book discusses quarks and leptons, antiparticles and Feynman diagrams, curved space-time, the Big Bang and the expanding Universe. Suitable for undergraduate students in non-science as well as science subjects, it uses problems and worked examples to help readers develop an understanding of what recent advances in physics actually mean"-- Provided by publisher.

Contents	Machine generated contents note: 1. Classical foundations; 2. Special relativity; 3. Quantum mechanics; 4. Elementary particles; 5. Cosmology.
Subjects	Quantum theory--Popular works. General relativity (Physics)--Popular works. Science / Cosmology.
Notes	Includes index.

Scalar boson decays to tau leptons.

LCCN	2017957836
Type of material	Book
Main title	Scalar boson decays to tau leptons.
Published/Produced	New York, NY: Springer Berlin Heidelberg, 2017.
ISBN	9783319706498

TAU2012: the Twelfth International Workshop on Tau-Lepton Physics: Nagoya, Japan, 17-21 September 2012

LCCN	2015490301
Type of material	Book
Meeting name	TAU (International Workshop on Tau Lepton Physics) (12th: 2012: Nagoya-shi, Japan)
Main title	TAU2012: the Twelfth International Workshop on Tau-Lepton Physics: Nagoya, Japan, 17-21 September 2012 / edited by Kiyoshi Hayasaka, Toru Iijima, Nagoya University, Japan.
Published/Produced	Amsterdam: Elsevier, [2014]
Description	xi, 239 pages: illustrations; 27 cm.
LC classification	QC793.5.L42 T36 2012
Variant title	Tau 2012
Related names	Hayasaka, Kiyoshi, editor.
	Iijima, Toru, editor.
Subjects	Leptons (Nuclear physics)--Congresses.
Notes	Includes bibliographical references and index.
Series	Nuclear physics. B, Proc. suppl., 0920-5632; 253-255

Tau2014: the 13th International Workshop on Tau Lepton Physics: RWTH Aachen University, Germany, 15-19 September 2014

LCCN	2015452373
Type of material	Book
Meeting name	TAU (International Workshop on Tau Lepton Physics) (13th: 2014: Aachen, Germany)
Main title	Tau2014: the 13th International Workshop on Tau Lepton Physics: RWTH Aachen University, Germany, 15-19 September 2014 / edited by Prof. Dr. Achim Stahl, Dr. Ian M. Nugent.
Published/Produced	Amsterdam: Elsevier, [2015]
Description	xv, 263 pages: illustrations; 27 cm.
LC classification	QC793.5.L42 T36 2014

Variant title	Tau 2014
Related names	Stahl, Achim, 1962- editor.
	Nugent, Ian M., editor.
Subjects	Leptons (Nuclear physics)--Congresses.
Notes	Includes bibliographical references and author index.
Series	Nuclear and particle physics proceedings, 2405-6014; volume 260

Tau2016: the 14th International Workshop on Tau Lepton Physics: Institute of High Energy Physics, Beijing, China, 19-23 September 2016

LCCN	2017470952
Type of material	Book
Meeting name	TAU (International Workshop on Tau Lepton Physics) (14th: 2016: Beijing, China)
Main title	Tau2016: the 14th International Workshop on Tau Lepton Physics: Institute of High Energy Physics, Beijing, China, 19-23 September 2016 / edited by Prof. Changzheng Yuan, Prof. Xiaohu Mo, Dr. Liangliang Wang.
Published/Produced	Amsterdam: Elsevier, [2017]
Description	xv, 227 pages: illustrations; 27 cm.
LC classification	QC793.5.L42 T36 2016
Variant title	Tau 2016
Related names	Yuan, Changzheng, editor.
	Mo, Xiaohu, editor.
	Wang, Liangliang (Physicist), editor.
Subjects	Leptons (Nuclear physics)--Congresses.
Notes	Includes bibliographical references and author index.
Series	Nuclear and particle physics proceedings, 2405-6014; 287-288

The standard model in a nutshell

LCCN	2016040024
Type of material	Book
Personal name	Goldberg, Dave, 1974- author.
Main title	The standard model in a nutshell / Dave Goldberg.
Published/Produced	Princeton; Oxford: Princeton University Press, [2017] ©2017
Description	xvii, 295 pages: illustrations; 27 cm.
Links	Publisher description https://www.loc.gov/catdir/enhancements/fy1618/2016040024-d.html Contributor biographical information https://www.loc.gov/catdir/enhancements/fy1701/2016040024-b.html
ISBN	9780691167596 (hardcover; alk. paper) 0691167591 (hardcover; alk. paper)
LC classification	QC794.6.S75 G65 2017
Contents	Special relativity -- Scalar fields -- Noether's theorem -- Symmetry -- The Dirac equation -- Electromagnetism -- Quantum electrodynamics -- The weak interaction -- Electroweak unification -- Leptons and quarks -- Particle mixing -- The strong interaction -- Beyond the standard model.
Subjects	Standard model (Nuclear physics) Particles (Nuclear physics) Symmetry (Physics)
Notes	Includes bibliographical references (pages 283-289) and index.
Series	In a nutshell In a nutshell (Princeton, N.J.)

Understanding Higgs bosons

LCCN	2015010702
Type of material	Book
Personal name	Bortz, Fred, 1944- author.
Main title	Understanding Higgs bosons / Fred Bortz.
Published/Produced	New York: Cavendish Square Publishing, [2016] ©2016
Description	64 pages: color illustrations; 24 cm
ISBN	9781502605504 (library bound)
	9781502605511 (ebook)
LC classification	QC793.5.B62 B668 2016
Contents	Inside the atom -- Discovering the particle "zoo" -- Quarks, leptons, and bosons -- The Higgs boson and beyond.
Subjects	Higgs bosons--Juvenile literature.
	Particles (Nuclear physics)--Juvenile literature.
Notes	Includes bibliographical references (pages 58-59) and index.
	7-12.
Series	Exploring the subatomic world
	Exploring the subatomic world.

Related Nova Publications

Fundamental Leptons as Compositions of Massless Preons: An Alternative to Higgs Mechanism[*]

Yu. P. Goncharov[†]

Theoretical Group, Experimental Physics Department,
State Polytechnical University,
Sankt-Petersburg, Russia

Within the framework of the confinement mechanism proposed earlier by the author in quantum chromodynamics (QCD) the problem of masses for fundamental leptons in particle physics is discussed for muon and -lepton. It is shown that the observed parameters of the mentioned leptons

---

[*] The full version of this chapter can be found in *Horizons in World Physics. Volume 285*, edited by Albert Reimer, published by Nova Science Publishers, Inc, New York, 2015.
[†] Email address: ygonch77@yandex.ru

such as their masses and magnetic moments can be obtained in a preon model dynamically due to a preon gauge interaction. The radii of fundamental leptons are also estimated. Under the circumstances preons might be massless in virtue of existence of the nonzero chiral limit for the preon interaction energy.

SELF-INTERACTION MASS FORMULA THAT RELATES ALL LEPTONS AND QUARKS TO THE ELECTRON[*]

Gerald Rosen[†]

Department of Physics,
Drexel University,
Philadelphia, PA, US

An accurate empirical self-interaction mass formula has been obtained for all twelve leptons and quarks, in which the baryon number B and the charge number Q enter as eigenvalue surrogates for the strong and the electromagnetic self-interactions. All lepton and quark masses appear directly related to the electron mass.

---

[*] The full version of this chapter can be found in *Recent Advances in Quarks Research*, edited by Haruki Fujikage and Kyou Hyobanshi, published by Nova Science Publishers, Inc, New York, 2012.
[†] Email: gerald.h.rosen@drexel.edu.

DIRAC MODEL EXTENSION FOR FINITE-SIZE LEPTONS AND QUARKS IN (10+1)D SPACETIME: QUANTUM CONDITIONS THAT IMPLY $A^{-1}=137.0360824$[*]

Gerald Rosen[†]

Department of Physics,
Drexel University,
Philadelphia, PA, US

In a Dirac model extension to 10 spatial dimensions (3 external plus 7 internal), leptons and quarks at rest are viewed to have a universal size and prolate spheroid shape, with mass and charge differences wholly attributed to their energy and charge densities through the universal spheroid. It is noted that simple volumetric quantum conditions on the 10D prolate spheroid and on its 4D prolate spheroid projection imply the fine-structure constant theoretical value $\alpha = (137.0360824)^{-1}$.

---

[*] The full version of this chapter can be found in *Recent Advances in Quarks Research*, edited by Haruki Fujikage and Kyou Hyobanshi, published by Nova Science Publishers, Inc, New York, 2012.
[†] Email: gerald.h.rosen@drexel.edu.

INDEX

A

amplitude, 61, 63, 64
annihilation, 21, 101
antineutrinos, 58
assets, 100, 106
astrophysical experiments, vii, viii, 57, 67
asymmetry, 97
atoms, 97

B

baryon, 79, 81, 97, 103, 114
Big Bang, 105, 107
boson(s), vii, viii, 2, 25, 28, 50, 51, 57, 58, 61, 62, 63, 64, 67, 69, 72, 73, 74, 75, 76, 80, 82, 86, 89, 91, 94, 96, 99, 100, 101, 102, 104, 105, 107, 111
bounds, 70, 72, 97

C

CDM, 21, 22
CERN, 2, 54, 88, 99, 100, 102
classification, 79, 81, 83, 84, 85, 86, 87, 88, 90, 91, 93, 94, 96, 98, 100, 102, 105, 106, 108, 109, 110, 111
Clifford algebra, 29, 54
cold dark matter, 28
cosmic rays, 21, 22
coupling constants, 2, 67, 75

D

dark matter, viii, 1, 18, 21
decay, 20, 28, 29, 73, 74, 75, 85, 95, 97, 99, 104
decomposition, 90
decoupling, viii, 57, 61, 67
dense matter, 58
dipole moments, 58, 72
Dirac equation, 33, 34, 35, 36, 101, 110
dispersion, vii, viii, 67

E

ECS-Hamiltonian quaternions, vii, viii, 27, 51
ECS-rotational SO(3) group, vii, viii, 27, 29, 51
effective field theory, 86

Index

electric charge, 3, 10, 13, 15, 23, 29, 37, 39, 40, 55, 59, 71
electric field, 23
electromagnetic, 4, 58, 59, 60, 61, 67, 70, 72, 74, 85, 91, 92, 97, 114
electromagnetism, 51
electron, 3, 5, 6, 8, 9, 10, 14, 21, 22, 23, 42, 58, 60, 66, 67, 71, 74, 95, 114
electroweak interaction, 85, 96, 102
elementary fermions masses, 1
elementary particle, 2, 6, 58, 69, 85, 88, 93, 98, 107
energy, 3, 4, 7, 10, 20, 23, 28, 30, 41, 51, 58, 59, 66, 67, 70, 75, 84, 86, 91, 98, 99, 102, 114, 115

F

fermions, vii, viii, 1, 2, 3, 18, 22, 28, 32, 61, 62, 72, 73
Feynman diagrams, 60, 71, 72, 73, 74, 85, 89, 107
field theory, 86, 90
fundamental forces, 80, 82

G

galaxies, 18, 93
gauge group, 28, 41, 51, 59
gauge invariant, 70
grand unified theories, 28, 52
grand unified theory, vii, viii, 27, 53
group theory, 28

H

hadrons, 80, 82, 87, 88, 95
Hamiltonian, vii, viii, 27, 41, 51, 90
Higgs boson, 2, 86, 89, 91, 94, 96, 98, 99, 100, 101, 102, 104, 105, 111

Higgs Field, v, 1, 2
hypothetical particles, 28

I

isospin, 38, 39, 100

L

Large Hadron Collider, 41, 91, 102, 103, 104
lepton, viii, 15, 28, 37, 38, 39, 40, 41, 45, 50, 69, 70, 71, 74, 76, 95, 96, 100, 113, 114
light, 20, 58, 89, 90, 97, 103
liquid phase, 80, 82

M

magnetic moment, vii, viii, 32, 57, 58, 60, 69, 70, 114
magnitude, 4, 6
mass, vii, viii, 2, 3, 4, 5, 6, 9, 10, 11, 13, 15, 16, 17, 18, 19, 20, 21, 22, 24, 28, 32, 35, 36, 37, 38, 39, 57, 58, 60, 67, 69, 71, 74, 75, 76, 80, 82, 95, 96, 103, 114, 115
massive particles, 21
mathematics, 83, 106
matrix, 33, 48, 55, 64, 97
matter, 18, 30, 58, 69, 70, 84, 93, 95, 97, 100
mesons, 29, 79, 81, 93, 95
Milky Way, 25
mixing, vii, viii, 11, 12, 57, 61, 62, 67, 69, 73, 76, 80, 82, 86, 96, 100, 102, 110
models, 59, 86, 87, 97, 104
momentum, 3, 4, 7, 10, 22, 30, 60, 71, 75, 80, 82, 100
muon collider, 103
muons, 24

Index

N

neutrinos, viii, 1, 2, 7, 10, 11, 12, 18, 20, 21, 22, 24, 58, 61, 96, 97, 102, 103
nuclei, 84, 85, 87, 88

O

oscillation, 58, 97

P

partial differential equations, 35
particle mass, 5, 23, 59
particle physics, 80, 81, 84, 87, 88, 91, 98, 100, 105, 106, 109, 113
photons, viii, 4, 6, 7, 8, 9, 13, 23, 28, 51, 89, 93
polarization, 23, 74
positron(s), 21, 23, 101

Q

quantum chromodynamics (QCD), 14, 24, 80, 81, 85, 86, 90, 91, 99, 113
quantum electrodynamics (QED), 85, 91
quantum field theory, 70
quantum mechanics, 88, 107
quantum state, 7, 12
quantum theory, 7, 13, 22, 98
quarks, 2, 13, 14, 16, 19, 24, 70, 79, 80, 81, 84, 85, 86, 89, 90, 91, 94, 95, 100, 102, 105, 106, 107, 110, 114, 115

R

radiation, 18, 97
radius, vii, viii, 3, 4, 5, 6, 30, 31, 57, 60, 67, 71
reactions, 59, 60, 66, 87, 92, 106
rotations, vii, viii, 28, 40, 48, 51

S

scattering, 58, 60, 80, 82, 101, 104
space-time, 29, 30, 31, 32, 33, 37, 41, 107
spin, 32, 34, 61, 63, 71, 98, 99, 103
Standard Model, viii, 25, 28, 57, 58, 67, 69, 84, 86, 95, 98, 99, 100, 101
structure, 3, 21, 58, 70, 72, 80, 82, 84, 91, 115
supersymmetry, 52, 80, 82, 102, 103
symmetry, vii, viii, 27, 28, 29, 37, 40, 41, 51, 70, 85, 90, 91, 95, 97, 99, 104

T

tau, 2, 3, 6, 8, 9, 12, 15, 23, 24, 37, 39, 71, 74, 75, 76, 104, 107
transformations, 28, 29, 39, 40

U

unification, 28, 80, 82, 85, 96, 99, 100, 101, 103, 110
universe, 88, 93, 96, 97, 104

V

vector, 4, 13, 40, 42, 47, 49, 64, 104

W

weak interaction, vii, viii, 27, 29, 51, 85, 101, 110

Polarons: Recent Progress and Perspectives

Editor: Amel Laref (Physics department, College of Science, King Saud University, Riyadh, Saudi Arabia)

Series: Physics Research and Technology

Book Description: This book presents recent research results on the illustrious verge of polaron science, which is broadly applied in condensed matter physics, solid state physics, and chemistry fields. It covers the modern progress of the polaron effect in various classes of materials.

Hardcover ISBN: 978-1-53613-935-8
Retail Price: $310

Quark Matter: From Subquarks to the Universe

Author: Hidezumi Terazawa (Midlands Academy of Business and Technology)

Series: Physics Research and Technology

Book Description: The meaning of "quark matter" is twofold: It refers to 1) compound states of "subquarks" (the most fundamental constituents of matter), which quarks consist of, as "nuclear matter" to those of "nucleons" (the constituents of the nucleus), and 2) compound states of quarks that consist of roughly equal numbers of up, down, and strange quarks, and which may be absolutely stable.

Softcover ISBN: 978-1-53614-151-1
Retail Price: $82

Proceedings of the 2017 International Conference on "Physics, Mechanics of New Materials and Their Applications"

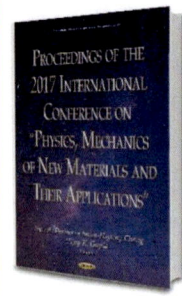

Editors: Ivan A. Parinov (Southern Federal University, Rostov-on-Don, Russia); Shun-Hsyung Chang, Ph.D. (Southern Federal University, Nanzih District, Kaohsiung City, Taiwan); Vijay K. Gupta (PDPM Indian Institute of Information Technology, Design and Manufacturing, Professor, Madhya Pradesh, India)

Series: Physics Research and Technology

Book Description: The book presents new results of internationally recognized scientific teams in the fields of materials science, physics, mechanics, manufacturing techniques and technologies of advanced materials, operating in diapasons from the nanometer level to the macroscopic level.

Hardcover ISBN: 978-1-53614-083-5
Retail Price: $310

Proceedings of the 2016 International Conference on "Physics, Mechanics of New Materials and Their Applications"

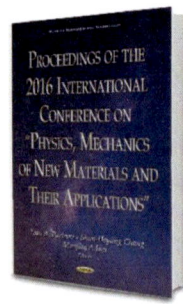

Editors: Ivan A. Parinov (Vorovich Mechanics and Applied Mathematics Research Institute, Southern Federal University, Rostov-on-Don, Russia); Shun-Hsyung Chang, Ph.D. (Department of Microelectronic Engineering, National Kaohsuing University of Science and Technology, Kaohsuing, Taiwan); Muaffaq A. Jani (PDPM Indian Institute of Information Technology, Design and Manufacturing, Madhya Pradesh, India)

Series: Physics Research and Technology

Book Description: The book covers broad classes of modern materials, structures and composites with specific properties. It presents nanotechnology approaches, modern piezoelectric techniques, physical and mechanical studies of the structure-sensitive properties of the materials, modern methods and techniques of physical experiment, etc.

Hardcover ISBN: 978-1-53611-033-3
Retail Price: $270